WHAT YOUR COLLEAGUES ARE SAYING . . .

One of the saddest comments we often hear is "I was never any good at math." People blame themselves or the math. Rarely do they blame the mismatch between their cognitive and emotional needs and how they were being taught. In this engaging book, Lidia Gonzalez shines a light on the cultural, curricular, and classroom realities that are the real culprits.

—**Steve Leinwand**
American Institutes for Research
Washington, DC

Insight into why we need to change the narrative, "I'm bad at math!" So many moments of "*Yes*, you hit the nail on the head!" Authentic stories and compelling evidence reveal how our society continues to perpetuate this harmful myth. There are abundant resources to help stakeholders dismantle systemic barriers that persist in math and math education and reflection questions for education professionals. Awesome work!

—**Shelly M. Jones**
Professor, Mathematics Education,
Central Connecticut State University
Hamden, CT

Bad at Math? creates the space to unpack people's dispositions about mathematics. Many people dislike how mathematics is used to position them as either competent or incompetent. This book provides the content and context for people to unpack mathematics as the tool that helps us critique and understand the world.

—**Robert Q. Berry III**
Dean and Professor, College of Education,
University of Arizona
Tucson, AZ

This book was a pleasure to read and reread! Though the main discussion is mathematics, it should be a must-read for all preschool through higher education professionals. It's well written, and deeply rooted research tells the story. The long overdue, honest discussion is chock full of inclusive history and timely strategies positioning us to move forward and do better!

—**Michele R. Dean**
Field Placement Director and Lecturer, Graduate School of Education,
California Lutheran University
Thousand Oaks, CA

What a powerful and thought-provoking book! Gonzalez does a masterful job of addressing what is wrong with mathematics education currently and what can be done to make mathematics more accessible for more students, particularly those who are marginalized. Through these changes, we can help make it less socially acceptable for people to say they are bad at math.

—Kevin J. Dykema
President, National Council of Teachers of Mathematics (2022–2024);
Eighth-Grade Mathematics Teacher, Mattawan Middle School
Mattawan, MI

This book truly breaks down cultural norms to build up a powerful vision of mathematics for everyone. The engaging and thought-provoking discussions are paired with rich examples and resources that collectively create a powerful message to help us change the way math is perceived and achieved in schools. An important book for all education stakeholders!

—Jennifer Bay-Williams
Professor, University of Louisville
Pewee Valley, KY

Bad at Math?

To my daughter Sofia—you are the light of my life—
and to my father whose light, unfortunately, went out too soon.

Bad at Math?

Dismantling Harmful Beliefs That Hinder Equitable Mathematics Education

Lidia Gonzalez

CORWIN Mathematics

FOR INFORMATION:

Corwin

A SAGE Company

2455 Teller Road

Thousand Oaks, California 91320

(800) 233-9936

www.corwin.com

SAGE Publications Ltd.

1 Oliver's Yard

55 City Road

London EC1Y 1SP

United Kingdom

SAGE Publications India Pvt. Ltd.

B 1/I 1 Mohan Cooperative Industrial Area

Mathura Road, New Delhi 110 044

India

SAGE Publications Asia-Pacific Pte. Ltd.

18 Cross Street #10-10/11/12

China Square Central

Singapore 048423

President: Mike Soules

Vice President and
 Editorial Director: Monica Eckman

Associate Director and
 Publisher, STEM: Erin Null

Content Development
 Editor: Jessica Vidal

Senior Editorial Assistant: Nyle De Leon

Production Editor: Tori Mirsadjadi

Copy Editor: Melinda Masson

Typesetter: C&M Digitals (P) Ltd.

Proofreader: Talia Greenberg

Cover Designer: Scott Van Atta

Marketing Manager: Margaret O'Connor

Printed in Canada

ISBN 978-1-0718-8717-2

This book is printed on acid-free paper.

23 24 25 26 27 10 9 8 7 6 5 4 3 2 1

CONTENTS

 You can access a Book Study Guide for *Bad at Math?* at
resources.corwin.com/badatmath

ACKNOWLEDGMENTS

First and foremost, I would like to thank the reviewers who read early drafts of this work and whose comments, feedback, and insight led to a clearer, more focused, and hopefully more useful final draft. Many thanks to my editor Erin Null, a publisher at Corwin Press, who saw promise in this book and whose guidance, feedback, and insightful questions helped move it along. When we first met, you indicated that you have published over 70 books about math education while at Corwin, and I am honored that mine is now among them. Thanks also to all the individuals at Corwin Press without whose assistance this book would not have been possible. Specifically, thanks to Nyle De Leon, Jessica Vidal, Melinda Masson, and Margaret O'Connor for their assistance throughout this process.

I'd like to extend thanks also to my mother, Luisa Gonzalez, for her constant reassurance that this book would be completed and for her encouragement now and always. Thanks also to my sister, Diana Siassos, a practicing elementary school teacher and the first person to read a draft of this work. Thanks to my daughter, Sofia, who is both an inspiration and a constant reminder that we need to get this thing we call mathematics education right for the next generation of students. Lastly, I want to thank my partner, Lina Del Plato, for her belief in me, her never-ending support, and her patience with me as I spent hours upon hours thinking, researching, and writing. My life and my work are better because of you.

PUBLISHER'S ACKNOWLEDGMENTS

Corwin gratefully acknowledges the contributions of the following reviewers:

Avital Amar
Math Consultant, York Region District School Board
Aurora, ON, Canada

Crystal Lancour
Supervisor of Curriculum and Instruction, Colonial School District
Middletown, DE

Casey McCormick
Math Teacher (Grades 5–8), Our Lady of the Assumption School
Citrus Heights, CA

Jacqueline Mickle
Elementary Math Consultant
Burlington, ON, Canada

Nicholas Pyzik
Elementary Mathematics Resource Teacher, Baltimore County Public Schools
Hampstead, MD

Stephanie Woldum
High School Math Teacher, Minneapolis South High School
Minneapolis, MN

ABOUT THE AUTHOR

Lidia Gonzalez is a professor in the Department of Mathematics and Computer Science at York College of the City University of New York. A first-generation college graduate, she began her career as a high school mathematics teacher in a large, comprehensive high school in New York City. Interested in improving the mathematical experiences of urban students, she focuses her research on the teaching of mathematics for social justice, the development of mathematics identity, and teacher development. She has published numerous articles and book chapters, has given talks at regional and national conferences, and is the recipient of several research grants. If you are interested in having Dr. Gonzalez come to your school to give a talk, to run an extended workshop, or to craft a customized professional learning experience, reach out to her at lgonzalez@york.cuny.edu

BAD AT MATH? AN INTRODUCTION

The most common response I get when I tell someone that I teach math for a living is that they were never good at math. There are variations to this answer, of course, but the essential point is that the individual I am speaking to is *bad at math.* Have you heard this comment or even said it yourself at times? Personally, I have gotten this response throughout my entire adult life both from those who have had little formal schooling and from those who have terminal degrees in their fields. As a high school teacher, when speaking to parents about the difficulties their child was having in mathematics, I was frequently surprised by their admission that they too had trouble with math. Now, as a college professor, I'm told by students and other faculty alike that they are *bad at math.* Among my colleagues, this is striking and perhaps unexpected. We are people who have devoted our lives to the pursuit of knowledge, yet, even among these learned individuals, I hear the *bad at math* comment frequently. Let me make this clear: Individuals who possess a PhD, who have published articles and books, who create new knowledge through research, and who are, by all the traditional measures, smart are comfortable saying that they are not good at mathematics. And they are not the only ones.

During the third quarter of Super Bowl LV, the Tampa Bay Buccaneers led the Kansas City Chiefs 24 to 9. The Buccaneers had possession of the ball and seemed poised to score again. One of the CBS announcers noted that he felt the Chiefs could come back at this point—being down by 15 points—but said that if Tampa Bay scored a touchdown or, perhaps, even two, this would no longer be true. Then he fumbled through the calculation of how many points behind the Chiefs would be should each of these scenarios occur. At one point he said, "I have trouble with the nines," referring to the fact that the calculations involved subtracting 9 points from the Buccaneers' total in each case. There were an estimated 96.4 million people viewing Super Bowl LV on CBS as he said this. The fact that so many people across all walks of life are comfortable publicly admitting they are *bad at math* doesn't sit right with me. Some might feel shame admitting that they cannot read, yet when it comes to mathematics, people openly admit their inability to do math seemingly without shame or hesitation.

As a mathematics educator, I would love for all individuals to appreciate and understand mathematics the way that I do. The feeling I get when I finally understand a problem I have been struggling with for weeks is quite addictive and one of the reasons why I decided to study math in the first place. I believe wholeheartedly that many more people can grow to both love and excel at mathematics than currently do. They too can experience the rush that comes from finally seeing a way through a problem that one has been working on. However, for this to be the case, the way in which we conceptualize and teach mathematics in our society needs to fundamentally change. So, too, must our acceptance of the socially constructed belief that it is permissible for many among us to be bad at the subject. The implications of society's acceptance of this belief and related beliefs are vast; they have a profound effect on mathematics education and, as a result, on our society as a whole.

Throughout the 20-plus years during which I have been teaching mathematics, many educational reforms have come and gone. The stated goals of adopting them have included increasing the number of students who excel in mathematics. These reforms have not truly been successful for myriad reasons. One is that the reforms have lived alongside the belief that to be bad at math is normal and to be expected. That is, there exists a crosscurrent that erodes efforts made to engage all students in the successful study of mathematics. Reforms, however well intentioned they may be, fall short when society fails to fully believe that they can be successful in a way that ensures that all (or even most) students excel in mathematics.

> **Reforms, however well intentioned they may be, fall short when society fails to fully believe that they can be successful in a way that ensures that all (or even most) students excel in mathematics.**

In their book, *The Stories We Tell: Math, Race, Bias and Opportunity*, Faulkner et al. (2019) talk about *belief stories* and their ability to influence our decisions even in the face of data that contradict the story itself. For example, if we believe that Black and Latinx students struggle with mathematics more than white students do, whether we admit this to ourselves or the belief operates more subconsciously, we will be less likely to refer Black and Latinx students to accelerated mathematics programs, even if the students we are selecting for such programs have similar academic records. Faulkner et al.'s work highlights the fact that blind referrals made without knowledge of the students' gender, race, and ethnicity lead to a greater number of Black and Latinx students being referred to advanced programs.

BELIEFS AROUND MATHEMATICS AND MATHEMATICS EDUCATION

The acceptance of failure in mathematics, just like all belief stories, permeates our society. It is perpetuated in the media and cemented in our popular culture. More troubling, however, is that this acceptance finds its way inside our classrooms, boardrooms, and government agencies. It impacts decisions around pedagogy, policy, and practice and affects the lives of those who must live with the consequences of such decisions. A further result of society's belief that it is okay to be *bad at math* is a narrowing of the conversation such that blame for failure is placed squarely on the individual. *I am bad at math*. Given this, the way to resolve the problem is for me—the individual—to receive tutoring, participate in a support program, dedicate more time to doing mathematics, or any number of other interventions. In framing difficulties in mathematics this way, we neglect the broader issues that impact mathematics education. We fail to consider the impact of a system of public education that is deprived of resources, one that disenfranchises students from marginalized communities, and one that often fails to support, value, and treat teachers like the professionals they are. We further neglect to push back against curricula that center algebra above all other branches of mathematics, textbooks that do not adequately reflect our students or value their lived experiences, and standardized exams that fail to adequately capture our students' abilities.

Similarly, we have, as a society, constructed other beliefs around mathematics and mathematics education that if not dismantled are harmful to the students we serve and the larger society of which we are a part. Here are some of those other dangerous beliefs:

1. Mathematics is all about numbers and equations.

2. Mathematics is about getting to the one correct—the *only* correct—answer.

3. Someone who does mathematics is smart, and part of what it means to be smart is to be able to do computations quickly in one's head without the need for aids or research.

4. There exist a small number of *math people* for whom mathematics comes naturally.

5. The educational system is somehow irreparably broken.

6. There exist achievement gaps in mathematics.

7. It is not important to attend to identity when teaching mathematics.

8. Mathematics is neutral and its teaching apolitical.

Each of these commonly held beliefs impacts the teaching and learning of mathematics. Further, they frame the discussion around mathematics education; they define teacher preparation programs; they are reflected in teacher licensing requirements; they inform the development of policies, funding, and curricula; and in the end, they have a broad and lasting effect on the teaching and learning of mathematics and the students we aim to serve. We need to acknowledge that the way we frame mathematics and mathematics education also forces upon us ways of responding, engaging, and reforming the discipline. Thus, efforts at meaningful and sustained change for the better require us to attend to these constructs.

Additionally, we cannot separate our discussion of the beliefs that frame mathematics and mathematics education from the society in which this education system is embedded. Dr. Jean Anyon (1997), an educational researcher who explored the inequities around schooling in U.S. society, put it very clearly when she said, "attempting to fix inner city schools without fixing the city in which they are embedded is like trying to clean the air on one side of a screen door" (p. 168). At this point in history, we can no longer deny that our society is built upon institutionalized racism, which fundamentally affects our system of schooling and thus the teaching and learning of mathematics. Additional forms of oppression such as sexism, ageism, ableism, heterosexism, and classism have impacted and continue to impact the development of the systems of public education that exist in many places in the world today—especially in the United States and Canada. These forms of oppression play a pivotal role in the lives of the students and families served by the school systems therein as well as the lives of the faculty, administrators, and staff who work in them. Any attempt to improve mathematics education must acknowledge the fact that our educational systems—from classroom interactions, to teacher preparation, to school funding, to curriculum—exist within societies that are rife with inequality and in which power and privilege play a prominent role.

> Any attempt to improve mathematics education
> must acknowledge the fact that our educational
> system—from classroom interactions, to teacher
> preparation, to school funding, to curriculum—exist
> within a society that is rife with inequality and in
> which power and privilege play a prominent role.

Therefore, any attempt to understand this system and dismantle the beliefs that drive the teaching and learning of mathematics must attend to these realities.

WHOM IS THIS BOOK FOR?

This book is written for all those with an interest in the teaching and learning of mathematics. Most especially, it is aimed at the teachers, administrators, and instructional leaders such as mathematics coaches, mentors, professional development providers, and teacher educators who work in schools today. You live the realities described herein and are uniquely positioned to lead efforts toward chipping away the harmful beliefs that currently exist—beliefs that, if not dismantled, severely limit efforts at improving the educational experiences of students with respect to mathematics. The ways in which you, as teachers, engage with students, parents, and the content you teach impact how mathematics is viewed and the beliefs that students and others cement around mathematics and mathematics education. As instructional leaders, you work with teachers to strengthen the ways that mathematics is taught. Your suggestions, your support, and the discussions you lead impact how these teachers conceive of mathematics, how they develop as educators, and what beliefs they pass on to their students. As administrators, you make decisions around curriculum, academic policies, budget, and hiring that impact the environment in which students, teachers, and instructional leaders work. It is this environment in which most people develop their beliefs about mathematics, and it is this environment where we can rewrite the story of what mathematics is, what it means to be good at it, and who can excel at it. You all, in your related and varied roles, have substantial influence on how mathematics will come to be seen and understood for generations to come.

HOW CAN THIS BOOK HELP?

The chapters that follow attempt to deconstruct commonly held beliefs about mathematics and/or mathematics education. Each chapter incorporates narrative and reflection into a discussion that highlights relevant research while paying particular attention to issues of power, privilege, and systems of oppression present in society. All of the chapters focus on the K–16 system in the United States and, to a lesser extent, Canada with a special emphasis on those schools that serve predominantly Black and Latinx students. They are also rooted in my experiences as an educator, a researcher, a student (the first in my family to go to college), a gay woman, the daughter of immigrants to the United States, and the parent of a school-aged child. Improving the mathematical experiences of those typically marginalized in mathematics is my passion and life's work. My hope, and the goal of this book, is that by critically examining the social constructs that frame mathematics and mathematics education, we can step outside the usual discourses, expand the conversation, and undertake authentic substantive changes toward equitable mathematics education.

This introduction is followed by 11 chapters. Chapter 1, *What Does It Mean to Be Good at Math?*, looks at the commonly held beliefs about what it means to be good at math and their implications. Chapter 2, *Beyond Numbers and Equations: What Is Mathematics?*, challenges traditionally held beliefs that center the definition of mathematics on numbers and algebra. In Chapter 3, *Mathematicians and Mathematicians in Training*, we examine commonly held beliefs about who mathematicians are and what they look like. We look at depictions of mathematicians in popular culture as places where stereotypes are reinforced. Chapter 4, *We Are All Math People*, confirms the existence of *math people* by redefining what that term means. In Chapter 5, *Identity in Mathematics Education*, we challenge the idea that mathematics education is such an objective discipline that it need not concern itself with issues of student identity. In Chapter 6, *School Mathematics*, we step back and uncover where many of our commonly held beliefs about mathematics and mathematics education originated: school. In Chapter 7, *Mathematics as Gatekeeper*, we examine the role that mathematics plays as a gatekeeper to future success. We look at mathematical testing and its role in our society. Chapters 8 and 9 move to system-level concerns surrounding education. In Chapter 8, *Achievement Gaps or Opportunity Gaps?*, we push back against commonly accepted narratives about achievement gaps between more affluent white students and their less mainstream peers. National discussions of the achievement gap are rampant within education, particularly as they relate to math and particularly in the wake of schools' reactions to the COVID-19 pandemic. In Chapter 9, *Is the School System Broken?*, we challenge commonly held beliefs about the purposes of schooling by considering the role that the educational system has as a reproducer of the inequality present in our social world. In Chapter 10, *Teaching Mathematics as a Political Act*, we challenge the commonly held idea that mathematics is neutral, objective, and apolitical. We consider the many ways that the teaching of mathematics is a political act in terms of what content is taught, whose stories are told, and how mathematics is contextualized. In the book's last chapter, *Where Do We Go From Here?*, we consider the power that you—as teachers, instructional leaders, and administrators—have to dismantle the harmful beliefs that currently exist around mathematics and mathematics education.

Each chapter ends with a series of questions for reflection aimed at teachers, instructional leaders, and administrators so that you can further engage with the ideas of the chapter. How can you, in your sphere of influence, take the ideas in the chapter and further them? How can you use your power, privilege, and position to act on the concepts therein to instill changes needed to strengthen mathematics education and best serve our most vulnerable students?

You will also find a Book Study Guide at resources.corwin.com/badatmath that is designed to help you and your colleagues work together as a community to digest the content in this book, try some of the activities together, and implement changes in your day-to-day practice.

I recognize that the topic of this book is complex, and the issues raised are beyond the influence of any one individual to fix. Only a collective effort will lead to needed change. I also recognize that I don't have all the answers—no one person does. For every facet of this topic, whole books could be written—and many have been! It would be impossible to cover it all in one single book, and I'm grateful to be able to stand upon and lean into the work of so many others striving collectively toward a more joyful and equitable mathematics learning experience for all. So, throughout this book, I try, where possible, to offer concrete ideas for small shifts you can start with, as well as additional resources such as books, websites, organizations, and podcasts to continue deepening your learning and practice.

HOW CAN YOU USE THIS BOOK?

This book should serve to ignite conversations at every level of education around what mathematics is, what it means to be good at mathematics, who is good at mathematics, and how the political and the mathematical are intertwined. It should serve as a springboard to talks about social justice and the ways mathematics education can promote it. It can be used in professional learning communities with in-service teachers as well as in courses for pre-service teachers. As the basis for professional development or a teacher book club read, this book has the potential to extend conversations already happening across pockets of North America. In the end, I call on you to build upon the work you already do to support equitable mathematics education and to join me in challenging the harmful beliefs that exist around mathematics and mathematics education so that there may come a day when no one is proud to announce they are *bad at math*.

A NOTE ON LANGUAGE

As Robin DiAngelo, best-selling author of *White Fragility* (2018), writes in her 2021 book *Nice Racism: How Progressive White People Perpetuate Racial Harm*, "Language is not neutral . . . the terms and phrases we use shape how we *perceive* or make meaning of what we observe" (p. xvii). As such, I acknowledge that the words I use in this text do not simply describe the realities discussed; they make meaning of them and reflect my own perceptions. Therefore, I feel it necessary before delving into the content of the book that I make clear some of my choices around language. When I am writing about racialized people, I use the current most recognized term for the group: *Black*, *Latinx*, and *Indigenous*. I do this even when the research I am citing differs in language (such as use of the term *Hispanic*) because I believe there is value in using terms that have evolved through time and are the current most recognized ones available. It is also in keeping with the current research in my field. I use the phrase *people of color* to describe all those who are nonwhite. I use the phrase *underrepresented groups in mathematics* in keeping with the definition of the National Science Foundation to describe those whose representation in mathematics degree programs and math-related careers is lower than their representation in the U.S. population (National Center for Science and Engineering Statistics, 2021). I use the phrase *students typically marginalized in mathematics* to describe not only those who are underrepresented but all who have been pushed to the margins of this discipline.

I often juxtapose the educational opportunities typically afforded to white students in contrast to those afforded to people of color, creating a binary dynamic that I understand fails to capture the many experiences and cultures present in both groups. This is, in part, due to the constraints of the research cited, of language in general, and of my own inability to find a way to address these issues while attending more adequately to the diversity within each group. I understand that collapsing peoples into these groups erases the nuances that exist within them, but hope that for the purposes herein these broad categories serve as useful and aligned with the current research literature. I capitalize *Black*, *Latinx*, and *Indigenous*. I once capitalized *white*, as it too is a racial category, but later learned that this is a practice among white supremacists. Not wanting to emulate such people, I no longer capitalize *white*.

When using terms like *men*, *women*, *girls*, and *boys*, I refer to all who identify with these terms—to people's primary identities. I acknowledge that some do not identify with either gender or do not do so consistently. I have avoided using *he* or *she* when gender is not known, choosing instead to use nongendered terms like *students*, *individuals*, and *people*, or the pronoun *they*. I could have added a third gender category but for brevity and flow did not. Some might be marginalized by this omission, and for that I do apologize.

I am hopeful that this clarifies some of the choices I have made with respect to language but acknowledge that no choice is perfect. I may be excluding some with my words while marginalizing others. Neither is intentional. I acknowledge too that I am part of the racialized system that exists in our society and that while I aim to be less racially oppressive, I too have been socialized into this system and play a role in the systemic racism that exists. I strive to be less ignorant on these matters, but my attempts are imperfect as well.

CHAPTER 1

WHAT DOES IT MEAN TO BE GOOD AT MATH?

In this chapter we will:

- Discuss and challenge existing beliefs about what it means to be good at math and how those beliefs commonly lead to the *bad at math* trope.

- Explore the role of open-ended problems in mathematics education.

- Consider the role of productive struggle and a growth mindset.

- Explore research on the role of brilliance in mathematics.

- Reflect on how you can challenge traditional views of what it means to be good at math.

WHAT BELIEFS EXIST AROUND WHAT IT MEANS TO BE GOOD AT MATH?

Society has developed a set of beliefs around what it means to be good at mathematics. For one thing, many believe that being good at mathematics means being able to complete numeric computations in one's head both quickly and accurately. Therefore, speed and computational skills are essential. Further, it is often believed that being good at mathematics means one can work through mathematical problems using previously memorized algorithms and procedures without additional aids such as a textbook or calculator—that is, that research and other tools are not necessary for those who excel in the subject. Our beliefs about what it means to be good at mathematics impact how the subject is taught. These beliefs also affect those who might wish to pursue its study, and they drive how we engage with the subject. The belief that to be good at mathematics one must be able to carry out numeric computations with speed and accuracy is driven by several factors, including

1. the types of problems most commonly featured in mathematics classrooms,

2. a bias toward algebra and number sense in our curriculum, and

3. representations of mathematics and mathematicians in the media and elsewhere.

Let us focus now on the first factor—the types of problems used in mathematics classrooms—and return to the others in later chapters.

TEXTBOOK PROBLEMS VERSUS OPEN PROBLEMS

There is a bias in mathematics textbooks and classrooms toward problems that are narrow in scope and that practice previously taught procedures. Students are taught a set of steps to solve problems of a certain type and then given a set of problems that test whether they have learned that procedure. This means there is an overreliance on procedures and less opportunity for the creative thinking that comes with exploring a more open-ended problem or one where the steps have not been given ahead of time. Mathematicians do not work on problems that have a set of steps previously laid out for them, so there is a need for creativity and approaching problems in multiple ways. The kinds of problems we typically see in classroom settings might be good at getting students to practice a particular algorithm, but they do not invite the type of open-ended problem solving that requires creativity, multiple approaches, and thoughtful study over a prolonged period. As a result, students fall victim to the false idea that they should know how to do math problems quickly.

There is a difference between the exercises we generally see in textbooks and true thoughtful, open-ended problems. We better serve our students by providing opportunities for both. Open-ended problems, as contrasted with closed problems, are those for which there is more than one solution and for which there are multiple methods to a solution. As an example, rather than ask what the sum of 10 and 5 is, one can ask the following related open-ended question: *The sum of two numbers is 15. What could the two numbers be?* Similarly, one can ask a student to prove, however they want to, that 55 is greater than 37. Here the student can subtract, draw pictures, compare the tens, plot both numbers on a number line, and so on. Lastly, instead of asking students to graph a line whose equation is given, one can ask them to provide the graph and equation of multiple lines through the point (1,2). To have them play with open-ended problems breaks down students' misconceptions about the nature of mathematics and builds up their ability in the subject by requiring them to use mathematics in creative and flexible ways. It also allows for questions to arise that might have otherwise not come to the surface. Open-middle problems are those that, while they have one solution, can be solved using multiple methods. These, too, have a place in our classrooms as they allow for creativity in problem solving. The idea is to use problems that lead to true authentic student thinking. As articulated in his text *Building Thinking Classrooms in Mathematics*, Peter Liljedahl (2021) spent over a decade studying the conditions and behaviors that create the optimal conditions for student thinking in mathematics classrooms. He makes a distinction between problems that promote mimicking—re-creating the steps a teacher has just gone through with a very similar problem—and those that promote actual thinking about mathematics. In the second case, students use mathematics and build mathematics in creative ways. They think about mathematics by looking for patterns, exploring connections, or extending the knowledge they have of the subject in some way. It is problems that promote this kind of thinking that often get neglected, and it is these problems that we should strive to bring into our teaching to help foster students' confidence and competence.

Additionally, utilizing complex problems without overly scaffolding them conveys that mathematics problems take time and creativity. Consider exploring the question of how many squares there are on a chessboard. A quickly obtained response might be 64, but this only counts the 1×1 squares. The whole board is a square, and there are many other squares in the board as well. To explore problems such as this normalizes the idea that some mathematical problems cannot be solved quickly or by simply applying an algorithm or calculation. Believing math problems can be solved quickly is harmful to students who, when faced with a problem they do not know how to solve, internalize the belief that they must be *bad at math*.

Recently, a parent shared with me another thing that for younger students may lead to misconceptions about the nature of mathematics: the amount of space given to solve various problems in workbooks and on worksheets. She noted that her children are often frustrated that even for problems that ask for an explanation of one's work only a small amount of space is given to work within. This might contribute to students' belief that the work and the thinking that leads to the solution to a mathematical problem should be short in nature. It might even reinforce the idea that what matters is an answer and not necessarily all the trials, work, and thoughts that get them to that answer—even when explicitly asked for—given how little space is offered to reflect their thinking. Perhaps encouraging students to use a blank piece of paper with more space or a whiteboard or journal would grow the belief that some mathematical problems might take time and much work to solve. Have you considered the amount of space your students have to work out mathematics?

Additionally, if we are honest with students about the fact that understanding comes with time and that multiple unsuccessful attempts often precede the attempt where one solves the problem, we will be teaching students that what matters is persistence and that an inability to solve a problem quickly is not proof of their inadequacy, but part of what it means to engage with mathematics. Students who struggle—or, worse, those who are labeled as struggling learners—internalize the belief that they are not good at math. Yet, struggle benefits us. Brain research demonstrates that our minds grow when we make mistakes and struggle (Boaler, 2015a; Moser et al., 2011). It is through work on tough conceptual problems that allow for us to get both stuck and unstuck that we grow as learners.

Struggle in mathematics is to be expected, and those who are successful are not those who do not encounter such struggle, but those who are able to persist despite it.

> **Struggle in mathematics is to be expected, and those who are successful are not those who do not encounter such struggle, but those who are able to persist despite it.**

To do mathematics is to be able to rethink one's approach, engage in research, try something new, regroup, and come at a problem in yet another way. It is a process known as *productive struggle*. Pasquale (2016) notes that when students struggle with a problem yet continue to make sense of it, they are engaging in productive struggle. Students who are most successful are those who can sit with the discomfort of not knowing and who are able to continue to work through problems regardless. Yet struggle is not what comes to mind when most people consider what it means to be good at mathematics.

I teach a mathematics course for future elementary school teachers in which I routinely have students work together in groups on an open-ended mathematics problem related to the topic for the day. These problems are not the typical, straightforward type that look the same as the one your professor just completed at the board but with different numbers. There is no set algorithm previously learned in class that can be applied to these problems. They require thought, creativity, and often multiple differing approaches before a group of students can see a way through to a solution. In the first few class sessions, students often take the problem and push it aside, claiming they don't know how to do this one or don't know the formula. They expect to be able to solve the problem quickly using a previously learned algorithm. If they cannot recall the algorithm, it must mean they cannot solve the problem.

I tell them that there is no formula and that they know enough to solve it, but that it requires them to *play* with the problem in much the same way that they would play with a puzzle, trying to fit a piece into various spots until they find where it belongs. Most are uncomfortable with the idea of playing in mathematics, though their comfort grows with each class meeting until they become comfortable at *play*. It takes time to begin to break down some of the ideas that students have about what it means to be good at mathematics, just as it will take time for society at large to rethink what it means to be good at mathematics. Nevertheless, this must occur.

There is a wonderful TED Talk, *Math Class Needs a Makeover*, by Dan Meyer (2010), in which he focuses on how we present mathematical content in mathematics classes in a way that provides for students all the information needed to solve a problem including visual representations, labels, symbolic language, and mathematical structure. We clean up the problems so much and create such small steps and substeps that students miss out on the opportunity to build mathematical understanding and to have mathematical conversations while considering authentic, rich problems. With respect to typical mass-produced textbooks, Meyer states, "What we are doing here is taking a compelling question, a compelling answer but we're paving a smooth, straight path from one to the other and congratulating our students for how well they can step over the small cracks in the way" (4:56). Instead, Meyer suggests teachers ask the shortest question possible and let the mathematics come out of the discussion that arises so that "math serves the conversation—the conversation doesn't serve math" (5:58). Instead of leading students through a discussion of slope, complete with the formula and notation needed, he models a situation where students are asked to consider what the steepest part of a ski lift is. By asking the shortest question, he begins a conversation that leads to a need for points, notation, symbols, and a way of defining steepness. The mathematics that comes out of that conversation is what you might expect to cover in a more typical lesson, except in this case the mathematics was developed, constructed, and built in conversation with students.

Meyer (2010) also challenges us to develop in our students the ability to be patient problem solvers. Typically, when students are asked to work on problems, they have a short amount of time to do so, yet complex mathematical problems take time to solve. There are starts and stops. You might try one approach and find the need to modify or abandon it altogether and try something new. You will often need to look up relevant research, see what others have tried, and note where the gaps are between what is already known and what needs to be known. There are problems that have taken centuries to solve and whose solutions have relied on the work of numerous mathematicians over the years. There are others that despite consistent and concentrated efforts over the course of decades, if not centuries, remain unsolved. In contrast to open-middle or open-ended problems, these are known simply as *open problems*.

One can, in fact, find lists of these open problems that have yet to be solved in mathematical journals or in books. There is a very well-known list of such problems put out by the Clay Mathematics Institute, a nonprofit organization dedicated to the mathematical research known as the millennium problems. The list of seven problems was released in 2000, and a $1 million prize is offered to anyone who solves one. The problems focus on a variety of subjects including number theory, algebraic geometry, and topology. An accessible explanation of the problems can be found at https://brilliant .org/wiki/millennium-prize-problems/. To date, only one of the millennium problems has been solved. It was known as the Poincaré conjecture. Proposed in 1904, it is about the topology of objects known as manifolds. It was solved by Grigori Perelman in 2006, a full 102 years after it was proposed. Interestingly, Perelman turned down the million-dollar prize and the Fields Medal, mathematics' highest award.

DEVELOPING RICH MATHEMATICS

Another difference between textbook exercises and open problems has to do with the mathematics that grows out of the work of solving these problems. It is common for new, valuable, rich mathematical ideas to be developed in the process of solving an open problem. While the solution to the problem might elude mathematicians for years, in working toward a solution, various techniques, methods, and mathematics are developed. Additionally, other related questions might arise for which solutions either are or are not able to be found. As an example, consider the Collatz conjecture, which is named after Lothar Collatz who put forth the idea in 1937. Begin with any counting number. If the number is even, you take half of it. If the number is odd, you multiply it by 3 and add 1. Now you have a new number and repeat the process. The chains for 5 and 7 are shown in Figure 1.1.

Figure 1.1 • *Collatz Conjecture Chains for 5 and 7*

Even? Divide by 2 Odd? Multiply by 3 and add 1	$5 \rightarrow 16 \rightarrow 8 \rightarrow 4 \rightarrow 2 \rightarrow 1$ $7 \rightarrow 22 \rightarrow 11 \rightarrow 34 \rightarrow 17 \rightarrow 52 \rightarrow 26 \rightarrow 13 \rightarrow 40 \rightarrow 20$ $\rightarrow 10 \rightarrow 5 \rightarrow 16 \rightarrow 8 \rightarrow 4 \rightarrow 2 \rightarrow 1$

The conjecture states that if we start with a counting number, *any* counting number, and apply these steps, we will eventually get the number 1. You might be wondering what happens if we do not stop at 1. In this case we get the chain in Figure 1.2, which loops forever.

Figure 1.2 • *Collatz Conjecture Chain for 1*

$1 \rightarrow 4 \rightarrow 2 \rightarrow 1 \rightarrow 4 \rightarrow 2 \rightarrow 1 \rightarrow 4 \rightarrow 2 \ldots$

No one knows for sure if every single number leads to a chain that eventually gets to 1, though many numbers have been checked. The fact that this conjecture remains open after over 80 years and yet is the subject of study has yielded interesting results and further questions for exploration. For example, questions have been asked about the length of the chains. The chains we created for 5 and 7 have six and seventeen numbers in them, respectively. The number 27, however, has 111 numbers in its chain. Which numbers yield very long chains, and which ones yield shorter chains? Chains of specific kinds of numbers have been investigated as well. For example, we can ask if there is anything interesting about the chains belonging to prime numbers. We can consider if there is anything interesting about the chains that belong to multiples of 7. The chain for 5 we created contains the number 8. We can ask what other chains contain the number 8. We can consider, too, questions around the frequency of even and odd numbers in certain chains. Rich problems such as the Collatz conjecture allow us not just to solve one problem, but also to discover, create, and see mathematical questions and avenues for further exploration. Even in this one example we can see how mathematics is alive with more and more questions unfolding before us. This example is given to highlight how an open problem, whether or not it is solved, leads to a plethora of new questions and avenues for exploration. These problems allow us to get lost in mathematics, following the questions as they arise in much the same way that one might get lost in a forest. In fact, getting lost in a mathematical forest is what Sunil Singh (2021) proposes to encourage mathematical wellness and inspire a love of the subject. Yet, while it might be nice to introduce a problem such as this so that students see the field as growing, students won't typically be working on problems that have gone unsolved for decades or more. They can, however, work on open-ended or open-middle problems with regularity, and doing so would benefit them.

Similar to how working on an open problem may yield new results and discoveries for a mathematician, students working on an open-ended problem (for which there may already exist a known solution) might discover mathematical patterns, use mathematics in ways that are new or innovative for them, and build their understanding of the field in a way not possible with a simple textbook exercise. Too few people have ever had the opportunity to work on open-ended problems, make mathematical discoveries, and build mathematics for themselves.

> **Too few people have ever had the opportunity to work on open-ended problems, make mathematical discoveries, and build mathematics for themselves.**

How might students' conceptions of what mathematics is differ if their experiences with the discipline involved the opportunity to ask questions and follow them on a quest of discovery?

MATHEMATICS AS AN EVOLVING FIELD

Unfortunately, many leave high school and college thinking of the field of mathematics as closed—that is, that mathematics was developed in the past and that there are no current, unsolved problems on which people are actually working. Mathematics is very much a living, creative, and evolving field. While much of what is taught at the undergraduate level, and prior, is centuries old and very well established, there are ways to highlight the evolving nature of mathematics, to talk about open problems, and to craft curriculum in such a way that allows students to see mathematics as filled with discovery, creativity, and imagination. For those who are interested, there are many journals, books, and online resources that provide open-ended, nontraditional creative problems for students and others to play around with and solve. As an example, the Association of Mathematics Teachers of New York State (AMTNYS) publishes the *New York State Mathematics Teachers' Journal* that features a problems section and accepts solutions for publication. It includes a set of problems called (*not so*) *elementary* where solutions are accepted only from elementary school teachers and precollege students. One of the most rewarding experiences I have had at the college level is working with two students who persisted through a problem posed in this journal. The students spent months working on the problem, trying new things, and coming at it with fresh ideas time and time again. Finally, they saw a way through, and the joy on their faces when they came to me with their plan for a solution—which they were certain would work—was wonderful, as was the excitement they brought with them the day they saw their solution published in the journal.

There is also a whole area of mathematics known as recreational mathematics, which gathers accessible yet exciting problems and topics for those who have an interest in them. Even a museum has been dedicated to bringing the joy of mathematics to the public. The National Museum of Mathematics opened its doors in New York City on December 15, 2012, and features exhibits that highlight the creativity, utility, and beauty of the subject in ways that are accessible to both children and adults. More information about the museum can be found at https://momath.org. MoMath, as it is known, also offers events one can participate in both in person and virtually, challenging the public to engage in mathematics in new and creative ways.

Resources: Recreational Mathematics

Averbach, B., & Chein, O. (2012). *Problem solving through recreational mathematics*. Courier.

Gardner, M. (2001). *The colossal book of mathematics: Classic puzzles, paradoxes, and problems: Number theory, algebra, geometry, probability, topology, game theory, infinity, and other topics of recreational mathematics*. W. W. Norton.

Gaskins, D. (n.d.). Math adventures for all ages. *Let's Play Math*. https://denisegaskins.com/internet-math-resources/math-adventures-for-all-ages/

Mathematical Association of America. (2022). *Recreational mathematics* [List of book reviews]. https://www.maa.org/tags/recreational-mathematics

O'Beirne, T. H. (2017). *Puzzles and paradoxes: Fascinating excursions in recreational mathematics*. Courier Dover.

Stewart, I. (2009). *Professor Stewart's cabinet of mathematical curiosities*. Basic Books.

DOES MATHEMATICS REQUIRE BRILLIANCE?

There is a related construct here, and that is the idea that being good at mathematics means having a natural talent or ability in the area that somehow cannot be developed through effort and a commitment to study alone. This is quite problematic, as it might keep some from pursuing the discipline while upholding an elitist view of mathematics as reserved for few instead of open to all. Yet, in a study of 1,800 academics who were asked to list the qualities needed to succeed in their discipline, mathematicians were among those who valued brilliance the most (Leslie et al., 2015). Brilliance is typically considered to be something that cannot be taught (Bian et al., 2018; Rattan et al., 2012). Further, mathematicians were more likely to attribute the success of women in their discipline to hard work, while attributing that of men to natural ability or brilliance (Leslie et al., 2015). Imagine sitting in a mathematics course where your instructor believes that to be successful in the field you need a natural ability, which that instructor feels cannot be taught.

> Imagine sitting in a mathematics course where your instructor believes that to be successful in the field you need a natural ability, which that instructor feels cannot be taught.

Next, imagine being a woman in this class who shows some success in the material but whose success is attributed by her instructor to effort alone and not to the natural brilliance the instructor feels is needed to be successful in the field. How supported do you think you might feel? How supportive might your instructors be of you if they held such beliefs? Yet this is what occurs, sadly, in far too many classrooms. The characterization that guides the beliefs of the instructors in this study goes beyond those instructors and their classrooms. These are not isolated cases, but rather part of a larger societal construct around how brilliance and mathematics connect. In their work, Meredith Meyer and her colleagues (2015) note that laypeople believe certain fields to require innate talent and that one of these fields is mathematics, so it is not just mathematics professors who hold this belief. This is one of the reasons why so many people feel comfortable admitting they are not good at it. Further, in Western societies, the belief exists that it is not possible for everyone to develop high intelligence. Aneeta Rattan, Krishna Savani, and their colleagues (2012) explored individual perceptions of the potential to develop high intelligence. They gave participants two related statements and asked them to choose which in each pair they agreed with most. As can be seen from Figure 1.3, over two thirds of those asked agreed with the statement that not everyone has the inborn potential to become highly intelligent, as opposed to the statement that everyone does have that potential, no matter how much they may desire to do so (Rattan, Savani, et al., 2012).

Figure 1.3 • *Perceptions of Potential for Intelligence*

Percentage of people who agree with each statement

Everyone has the inborn potential to be highly intelligent.	33%	67%	Not everyone has the inborn potential to be highly intelligent.
Everyone has the inborn potential to be highly intelligent if they want to.	51%	49%	Not everyone has the inborn potential to be highly intelligent if they want to.
Everyone has the inborn potential to be highly intelligent, but not all people end up realizing their potential.	49%	51%	Some people just don't have the inborn potential to be highly intelligent if they want to.

This viewpoint cements the idea that either you are good at mathematics or you are not, as there is no way to acquire the high intelligence that many believe is required for success in the field. All of this has implications for how students are, or are not, supported in mathematics learning and for society in general. We need to think about how to counter societal beliefs so that future generations see mathematics as more inclusive and so that we can support more diverse voices among its ranks.

The idea that to be good at math requires a natural brilliance that few have is supported, in part, by the media, which often use mathematics to highlight the intelligence of the characters portrayed or the difficulty of the work they are undertaking. When a student is struggling to do well on an exam and the camera moves from the character to the paper in front of them, it is most certainly a mathematics test they are working on. Mathematical symbols fill the chalkboard behind the emerging scientist who despite all odds, quickly and with certainty, manages to solve the equation that saves the day. Do you ever wonder why the exam is not in history or the chalkboard covered with musical notes?

If we believe brilliance is necessary for success in mathematics, that brilliance is limited to few, and that brilliance cannot be taught, it is difficult to imagine a reality where many decide to study mathematics in our society.

> **If we believe brilliance is necessary for success in mathematics, that brilliance is limited to few, and that brilliance cannot be taught, it is difficult to imagine a reality where many decide to study mathematics in our society.**

Further, brilliance, or natural talent, is most often associated with men. Consider that the phrase "Is my son gifted?" is googled two and a half times as often as "Is my daughter gifted?" (Stephens-Davidowitz, 2014). What does this say about the way we view our daughters? What does this say about our expectation that women can and will excel in mathematics? It is, again, hard to imagine that in a society in which these beliefs are prevalent, large numbers of women will be successful in mathematics.

Related to this is the idea that not everyone *can be* good at mathematics. This differs from the belief that not everyone *is* good at mathematics presently, but with effort, training, and the right support one can grow to be good at mathematics. To believe that one can grow to be good at mathematics is to believe that ability in the subject is not fixed. People who have this belief have what is called a growth mindset. A growth mindset allows for someone who might struggle presently with the subject to understand that they can with study and persistence improve their ability. Rattan, Good, and Dweck (2012) conducted a study of mathematics instructors at a competitive private university on the West Coast of the United States. They found that instructors who held a fixed mindset about mathematics were more likely to view those who scored below 65% on their first exam as *bad at math*. This is after just one exam. Keep in mind that these were students at a competitive private university, so they had to have already demonstrated some academic ability to have been admitted. Despite this, instructors with a fixed mindset not only were more likely to view students who scored below a 65% as *bad at math* but also expressed significantly lower expectations for these students when compared to faculty who had a growth mindset. These instructors were also "more likely to comfort students for their (presumed) low ability, and more likely to use teaching strategies that are less conducive to students' continued engagement with the field" (Rattan, Good, & Dweck, 2012, p. 734). By "comforting students," what the authors mean is that these instructors were more likely to tell students that their low ability was okay because not everyone can be good at math. These instructors were also more likely to counsel students out of the discipline and encourage them to study something else. Further, the students who received this feedback were fully aware of the fact that their instructors didn't believe in them and reported that their instructor was more likely to have lower engagement in their learning going forward in the course. Thus, these students weren't getting the support that they needed to do well in the subject, but they *were* getting the clear message that they were not expected to do well based on one exam alone. On the other hand, instructors with a growth mindset were more likely to employ strategy feedback with students who scored below a 65% on their first exam. These instructors were more likely to recommend tutoring, to talk to students about changing their study habits, and to commit to calling on them more in class and giving them more challenging tasks to support their learning. An extensive review of the research on teachers' and students' mindsets and their relationship to student achievement was conducted by Junfeng Zhang and colleagues at the University of Helsinki in 2017. They found that research conducted with school-aged children demonstrates that teacher mindset plays a role in the development of student mindset and as a result impacts student achievement in positive ways. Put simply, teachers with a growth mindset have the potential to positively impact their students at all grade levels.

The belief that mathematics is innate narrows the pool of those who feel comfortable pursuing it. Imagine the discoveries in mathematics that could be made if more people were involved in the development of the discipline. What if more diverse individuals were studying, crafting, and building mathematics? What could be accomplished then? That mathematics is the foundation of so many other disciplines and the basis for the technology on which we rely is well established. The better we are at encouraging more individuals and those from a wider array of backgrounds to study it, the better we will be able to grow and develop our technology and so much of what we depend on in society whether technological or not. Until we reframe what success in mathematics is and how one acquires it, we cannot change a system of mathematics education that privileges few and leaves out too many.

To explore what it means to be good at mathematics in the classroom:

- Have students write their thoughts about how one becomes good at something (including math), then open a conversation about perseverance, practice, and effort.

- Discuss specific skills for improving mathematics ability such as ways of studying and how to read or use a math textbook with your students.

- Share research on learning with your students.

- Remind students and yourself that everyone can learn math to a high degree.

- Give students the opportunity to revise their work, highlighting that what matters is learning over time.

WHAT CAN YOU DO TO CHALLENGE WHAT IT MEANS TO BE GOOD AT MATH?

Part of this work starts with rethinking how we define what it means to be good at mathematics. What a difference one's mindset about what it means to be good at mathematics can make. Let's strive to remind students, colleagues, families, and others we engage with that even if one does not excel in mathematics at present, all have the potential to, given the right supports.

There are several steps we can take to challenge the prevailing beliefs that exist around what it means to be good at mathematics, including engaging students in a discussion around this concept. Others involve

- uncovering our own biases,

- using rich problem-solving tasks,

- normalizing productive struggle, and

- promoting a growth mindset.

Uncovering Our Own Biases

We all have biases. Whether we admit that to ourselves or not, there are beliefs we hold that impact how we experience the world, how we interact with others, and how we engage in the teaching and learning of mathematics. We do well to try and unpack these biases so that we are more aware of them and able to consider how they impact our work. This is the first step in uncovering and nurturing our own strengths as educators (Berry, 2008). One way to start to uncover and interrupt our own biases around students' learning of mathematics is to look at students through a strengths-based lens (Kobett & Karp, 2020). Consider the statements in Figure 1.4, which may reflect things you have commonly heard or maybe even said. What is the underlying belief each statement might convey? How can reframing that statement through a strengths-based lens help turn the bias around?

As Beth Kobett and Karen Karp (2020) share, "The first step in changing the narrative is to consciously hear the language that we and others use to describe our students. When we work to identify a belief that focuses on moving in a positive direction, we are more likely to interact with other teachers, our students, and families in more positive and productive ways" (p. 28).

Now it's your turn to consider your beliefs about your students: Try visualizing a student who you think is good at math, and write down concrete reasons why you believe this to be true. What about a student who you think is struggling? What words have you used to describe this student in the past, and what might those statements convey about your underlying beliefs? What does this tell you about how you conceptualize what student success looks like in mathematics? How might you expand or challenge your reasons?

Figure 1.4 • *Shifting to Beliefs That Emphasize Students' Strengths*

Statement	Underlying Belief	Alternative Belief
She is doing the best she can.	She can't learn more.	She can learn math. We just need to find an entry point into her learning.
He can't help it.	He doesn't have self-control or self-regulation skills.	He has better self-control when he is able to select manipulatives, tools, and a place to work.
We can't expect her to do more.	She cannot learn more mathematics than she is currently learning. She is incapable of learning more.	If we raise our expectations and set success criteria in collaboration with the student, she will be able to achieve.
She lacks the background knowledge to grasp this information.	How much students can learn depends on the background knowledge they hold. Students are unable to learn without the right background knowledge.	She has solid knowledge about money. Let's use that knowledge to develop ideas about place value.
She does not care.	This student's behavior indicates that she does not value school.	The student's behavior indicates that we need to show her how much we care about her learning.
Even though he is motivated to learn, he is unable to retain the concepts.	Students who struggle with retention cannot learn mathematics.	I notice that he retains more when he is able to work with his peers to solve problems. Let's try pairing him with a classmate.
He can do this—he is just lazy.	Students choose to not work.	We need to find out why he does not complete his work.
You just need to tell them how to do it because they can't think on their own.	Direct instruction is best for students who struggle.	My students are capable of higher-level thinking and problem solving.
His parents don't care and can't help him.	Families that cannot attend conferences don't care about their child's learning.	Families care very much about their children's school success, but don't always show it in the same way or in ways that resonate with teachers' own families.

SOURCE: Kobett, B. M., & Karp, K. S. (2020). *Strengths-based teaching and learning in mathematics: Five teaching turnarounds for Grades K–6*. Corwin. Reprinted with permission.

Resources

- Check out the self-assessments in Chapter 1 of Kobett and Karp's text at resources.corwin.com/badatmath to learn about your mathematics identity as well as the beliefs you hold about yourself as a mathematics teacher and your students as learners of mathematics.

  - Kobett, B. M., & Karp, K. S. (2020). *Strengths-based teaching and learning in mathematics: Five teaching turnarounds for Grades K–6*. Corwin.

- Take Harvard's Project Implicit online assessment to identify attitudes, beliefs, and implicit biases around race, gender, and other areas.

  - Project Implicit. (2011). *Preliminary information: Implicit Association Test (IAT)*. https://implicit.harvard.edu/implicit/takeatest.html

- Write out what you think it means to be good at mathematics. Review your words and consider how your definition affects your work as a teacher, instructional leader, or administrator.

Using Rich Problem-Solving Tasks

One way to promote the view that being good at mathematics isn't about having an ability to compute quickly is to have students work on problems that are rich, complex, and open-ended, where not all the required symbolism, language, and steps are predetermined and have been previously taught. That is, you can have students build mathematics appropriate to their grade level by working on problems and activities that require them to do some research, employ some tools, and really see the creativity involved in the work. This does not have to be done exclusively, and certainly there is still a place for more traditional textbook problems, but too often these rich, open-ended problems are excluded from students' school experiences, especially for students in special education classes. Some believe that until students can master the textbook problems, they cannot engage in the higher-level thinking required for these more open-ended, rich problems, but the opposite is true. Until students have the opportunity to build mathematics and engage in rich open-ended problem solving, they may not solidify their understanding of the more traditional problems and may grow frustrated by a series of problems disconnected from themselves with preset steps they need to memorize rather than fully understand. They don't get at the *why*—why the method works— and they don't own the mathematics. They fail to see how a seemingly simple question can lead to an amazingly rich array of mathematics that they can have a hand in building with teacher guidance, but also with their own voices.

Further, when given a set of prescribed steps or guidance that overscaffolds a problem, students become less able to successfully work through problems that differ from those they are typically given. This is a case where less is more. Less scaffolding and less structure might afford students the opportunity to play with the mathematics in ways that are creative and allow them to think through the problem differently. In a study of two schools, one that used open-ended tasks in classes and one that used more traditional approaches, Boaler (2002, 2015a) found that in the school that used open-ended problems students who worked together toward a solution were better able to problem solve and to work through problems that differed from those they had previously seen. They were also more likely to say they enjoyed mathematics and years later were more likely to be in highly skilled and professional careers as compared to the students at the more traditional school.

Even the youngest of children can investigate patterns and draw inferences from these patterns. Students learning multiplication can look at multiples of 5 or 10 and draw conclusions about them, or they can investigate whether multiplying by an even number always yields an even number or whether multiplying by an odd always yields an odd and maybe start to get at why. Open-ended questions about the relative speed at which two vases of different shapes will fill when water is poured into them at the same rate can spark talks about volume, shape, rates, and measurements with little but a question posed. Students can be given the supplies needed to investigate these situations and begin to make predictions not just about which vase will fill first but about what properties matter and whether there is a way to tell in certain cases without much work. Exploring in which vase the water level rises fastest is a related problem that is worth considering as well. These questions and explorations are quite far removed from traditional textbook problems but can, in fact, teach the same concepts. One thing you can try in preparation for your next class is to rewrite a closed question you might have planned to use with your class as an open question.

Resources: Open-Ended Rich Problems

Dougherty, B., & Venenciano, L. (2023). *Classroom-ready rich algebra tasks, Grades 6–12: Engaging students in doing math.* Corwin.

Fletcher, G. (n.d.). *3 Act Task file cabinet.* https://gfletchy.com/3-act-lessons/

House, P., Stenglein, S., & Day, R. (2021). *Activity gems for the 9–12 classroom.* National Council of Teachers of Mathematics.

Kobett, B. M., Fennell, F., Karp, K. S., Andrews, D. R., Knighten, L. D., & Shih, J. C. (2021). *Classroom-ready rich math tasks, Grades K–1: Engaging students in doing math.* Corwin.

Kobett, B. M., Fennell, F., Karp, K. S., Andrews, D. R., & Mulroe, S. T. (2021). *Classroom-ready rich math tasks, Grades 4–5: Engaging students in doing math.* Corwin.

Kobett, B. M., Karp, K. S., Harrison, D., & Swartz, B. A. (2021). *Classroom-ready rich math tasks, Grades 2–3: Engaging students in doing math.* Corwin.

Open Middle Partnership. (2022). *Open Middle: Challenging math problems worth solving.* www.openmiddle.com

Orr, J., & Pearce, K. (Hosts). (2018–present). *Making math moments matter* [Audio podcast]. https://makemathmoments.com/podcast/

Smith, M., Bill, V., & Steele, M. D. (2020). *On your feet guide to modifying mathematical tasks.* Corwin.

Stanford Graduate School of Education. (n.d.). *YouCubed®: Inspire ALL students with open, creative, mindset mathematics.* Stanford University. www.youcubed.org

Normalizing Productive Struggle and Supporting a Growth Mindset

Let us normalize struggle in the classroom and beyond. It is often the case that when a student is struggling on a problem, our instinct is to help them by nudging them along through the work. I know I too am guilty of doing this, but it feeds into the idea that if one does not get a problem right away, something is wrong, and one needs help immediately. What if instead, as a classroom teacher, you gave students the time and space needed for them to struggle? What if as an instructional leader you promoted the idea that struggle is positive? What if it became the norm that we might have to try a problem more than once and in multiple ways to understand it?

If we normalized taking multiple approaches and at times multiple attempts to solve a problem, we would be building perseverance in our students, allowing them to engage in productive struggle and teaching them that what matters is not getting something quickly. Those students who are able to stick with the discomfort of not knowing and play with the math for longer despite it tend to do better, so why not cultivate this in students? I am suggesting not that we never step in to help but rather that we delay that help a bit in cases where we know that students—with more time, more approaches, and a willingness to stick with a problem—can in the end reach a solution if they believe that this will come with time, albeit not always with comfort. How much more resilient might students be if they realized that a stumble is just that and not a judgment on their abilities or a prediction of future failure?

One thing you can try in your next class is to ask a student who has solved a problem to solve it a different way and see what arises. The fact that mathematics problems can be tackled in so many ways yet reveal the same reality is amazing to me and something that we should value more in our teaching. When a student completes a problem and arrives at the right solution, we often end there. What if instead we asked that same student to consider a different path to the solution? How much more creative and versatile might their understandings be if students were pushed to look for multiple paths? In fact, mathematicians often do this. Perhaps they prove a particular theorem in some way. Then, they consider finding a more elegant proof, one that uses fewer steps, perhaps is more direct or more efficient in its approach, or has some other feature that is of value in some way even if that way is a matter of aesthetics. Develop in the budding mathematicians in your classes the ability to look at the same set of information in new ways: perhaps an algebraic approach followed by a geometric approach, or a numeric approach followed by a visual one. The Pythagorean theorem, as an example, has more distinct proofs than any theorem in history. In her 1968 text, *The Pythagorean Proposition*, Elisha Scott Loomis includes no less than 367 such proofs, including one by former U.S. president James Garfield. Being able to find another way is a valuable skill in life, so it seems useful to include this inclination as part of the way in which we teach mathematics.

Additionally, let us occasionally give students problems that we know will take a long time to solve, that will require multiple attempts, and that they perhaps will not be able to solve in the end. This will help normalize the idea that true mathematical work takes time and struggle and does not always yield a solution as we might have wanted. This is fine. It is the reality of the field and one students will benefit from knowing about. If we want people to fundamentally change the way they conceptualize who is good at mathematics, we must give students fundamentally different classroom experiences.

> If we want people to fundamentally change the way they conceptualize who is good at mathematics, we must give students fundamentally different classroom experiences.

One way to do this is to ensure that students have ample opportunities to work together on tasks and activities. Mathematicians often collaborate with others. In fact, conferences are set up with working groups where mathematicians of certain specialties gather to work on open problems in particular fields together. Developing mathematics is, in many cases, a social process. Let us ensure that our students have the opportunity to see and experience it as such.

Resources: Growth Mindset and Productive Struggle

Brock, A., & Hundley, H. (2018). *In other words: Phrases for growth mindset: A teacher's guide to empowering students through effective praise and feedback*. Simon & Schuster.

SanGiovanni, J. J., Katt, S., & Dykema, K. J. (2020). *Productive math struggle: A 6-point action plan for fostering perseverance*. Corwin.

In this chapter we considered people's perceptions around brilliance and its connection to mathematics. Here, we have much work to do. It is troubling to know that half of those asked believe that there are some individuals who lack the ability to do well in math and who despite their efforts cannot do well in it. This is flatly unacceptable. It erodes efforts at making the discipline one in which all can succeed, and it exemplifies a fixed mindset that is harmful to the people who hold these beliefs. We need extensive professional development around the idea of growth mindsets. Rather than counseling them out, we need to actively work with students who with some support would excel, and we need to ensure that more women and members of underrepresented groups in mathematics are supported rather than excluded not only by developing programs to support them but by actively changing the culture so that in time these programs are not needed. We must talk to students from an early age about growth mindsets and instill in them the belief that their ability to learn and grow is not fixed. You can do this both in and out of the classroom. When those around us talk in ways that support fixed mindsets or the idea that not all of us can excel in mathematics, we must challenge them with vigor. Remind them that "no one is born lacking the ability to learn math" (Boaler, 2015a, p. 5). Actively talk to your students about the ways in which they can grow as learners of mathematics. Consider also the feedback and praise you give to students and how your words reinforce or challenge the idea that intelligence is fixed. Praise is often used in classrooms as an effective tool, but it can reinforce fixed mindsets if we, for example, praise students for being *smart* or *having smart ideas*. Instead, try praising students for things they can control such as working hard on a task or trying multiple approaches. "Praise feels good, but when people are praised for who they are as a person ('You are so smart'), rather than what they did ('This is an amazing piece of work') they get the idea they have a fixed amount of ability" (Boaler, 2015a, p. 8). In the feedback you provide, you can model and support the idea that intelligence is something one can grow and develop over time.

Questions for Reflection

For Teachers

- What messages do you send to students about who is good at mathematics through your words, actions, the way you decorate your classroom, and the resources you employ?

- How might you bring open problems and the work of mathematicians into your teaching?

- What rich tasks can you employ to teach mathematical content?

- How can you talk to parents and others who may believe that it is socially acceptable to be bad at math?

- What stories did you believe about mathematics when you were a student? What stories do you believe now?

For Instructional Leaders

- In what ways do the activities you use with teachers promote the idea that to be good at math is to persist through problems?

- How do the resources you provide challenge traditional views of what it means to be good at math?

- How do you respond when someone admits to you that they are bad at math?

- How do you put into practice the idea of growth mindsets in your own work with teachers?

- What beliefs do you hold about what it means to be good at mathematics? How do these influence your work with teachers?

For Administrators

- How do the curriculum materials and resources your school adopts speak to what it means to be good at mathematics? How do they allow for productive struggle? How do they engage students in open-ended problem solving?

- What opportunities do you create for teachers, parents, and community members to grapple with the ideas in this chapter? Are there intentional spaces created for these individuals to come together to discuss ideas?

- How can you encourage productive struggle among your teachers by allowing them the space and time needed to try different teaching techniques and approaches to find what works best for them and their students?

- What beliefs do you hold about what it means to be good at mathematics? How do these affect your decisions around curriculum, hiring, and teacher development?

CHAPTER 2

BEYOND NUMBERS AND EQUATIONS

What Is Mathematics?

> In this chapter we will:
>
> - Consider what mathematics means and how it is used.
>
> - Explore research around student views of mathematics.
>
> - Illuminate important but underrepresented areas of mathematics.
>
> - Reflect on how you can challenge traditional views of what mathematics is.

WHAT COMES TO MIND WHEN YOU THINK OF MATHEMATICS?

One reason many people are comfortable saying they are *bad at math* is that they are thinking of mathematics in a very narrow way, focused on only those topics that they may have struggled with in school rather than on all the mathematics they undertake each day. Take a moment to consider the question *What is mathematics?* What ideas, words, and images jump to mind when you try to define the subject? For many, the answer revolves around numbers, computation, and equations. Images conjured are often of a chalkboard covered in symbolic language or of one's own experiences in a mathematics classroom. Less often do people think of their day-to-day lives and the mathematics that comprises so much of what they experience, or of the mathematics that underlies so many facets of our social, natural, and technological world. Yet, when you have a few errands to run, and you set out to determine the best route to take in completing them, you are engaging in mathematics. You may want to ensure you don't double back on any of the streets you have already traveled, if possible, to go about your errands more efficiently and save time. You might ask if it is possible to create a path that allows you to make it to each of your stops without crossing the same street more than once. These kinds of considerations are part of a field of math known as graph theory. Such questions have been studied for a long time.

The Seven Bridges of Königsberg

Figure 2.1 • *The Seven Bridges of Königsberg*

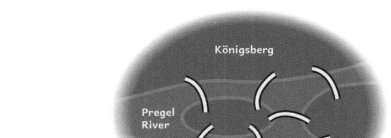

The city of Königsberg in Prussia (what is now Kaliningrad, Russia) comprises land on both sides of the Pregel River as well as two large islands (Kneiphof and Lonse) that were connected to each other and to the mainland portions of the city, as shown in Figure 2.1, by seven bridges. Those who lived there wondered whether it was possible to walk through the city by crossing each of the bridges once—and only once. Known as the Seven Bridges of Königsberg, the question was resolved in 1736 by mathematician Leonhard Euler, who determined it was impossible to walk over each of the bridges precisely one time. In doing so, he created the field of graph theory, an incredibly rich area of mathematics with applications to many societal realities, as we will discuss later in this chapter. Does this example fit into your conception of what mathematics is?

WHAT IS MATHEMATICS, REALLY?

Mathematics, when understood deeply, transcends numbers, equations, and computation. Keith Devlin, a Stanford University professor and National Public Radio's *Math Guy*, defines it as the study of patterns. Specifically, he writes:

> The mathematician identifies and analyzes abstract patterns—numerical patterns, patterns of shape, patterns of motion, patterns of behavior, voting patterns in a population, patterns of repeating chance events, and so on. Those patterns can be either real or imagined, visual or mental, static or dynamic, qualitative or quantitative, utilitarian or recreational. They can arise from the world around us, from the pursuit of science, or from the inner workings of the human mind. Different kinds of patterns give rise to different branches of mathematics. For example:
>
> - Arithmetic and number theory study the patterns of number and counting.
> - Geometry studies the patterns of shape.
> - Calculus allows us to handle the patterns of motion.
> - Logic studies patterns of reasoning.
> - Probability theory deals with patterns of chance.
> - Topology studies patterns of closeness and position.
> - Fractal geometry studies the self-similarity found in the natural world.
>
> (Devlin, 2012, p. 3)

Devlin notes further that in the mid-19th century mathematicians used mathematics to study mathematics itself and adopted a new conception of mathematics where the primary focus shifted away from calculation and computing answers to formulating an understanding of abstract concepts and relationships. "Mathematical objects were no longer thought of as given primarily by formulas, but rather as carriers of conceptual properties. Proving something was no longer a matter of transforming terms in accordance with rules, but a process of logical deduction from concepts" (Devlin, 2012, pp. 5–6). He refers to this new conception of mathematics as a revolution that "completely changed the way mathematicians thought of their subject" but adds that "for the rest of the world, the shift may as well not have occurred" (Devlin, 2012, p. 6). The reason for this is that the mathematics most people are exposed to in their schooling is very focused on computation and calculation. You are taught to identify types of problems and then given procedures to solve these problems. Devlin notes that most of the mathematics used today was developed in the last 200 years, but virtually none of what is taught is from even the past 300 years. As a result of this, many are "unlikely to appreciate that research in mathematics is a thriving, worldwide activity, or to accept that mathematics permeates, often to a considerable extent, most walks of present-day life and society" (Devlin, 2012, p. 2). I would argue and have already argued in this text that it does not need to be this way entirely. We can highlight the evolving nature of mathematics and talk about open problems in a way that allows individuals to see the creative and evolving nature of the field.

HOW DO STUDENTS VIEW MATHEMATICS?

A traditional view of mathematics as focused on numbers and equations can be found among students as well, even among those studying mathematics itself. One way to change this is to invite mathematicians and others in math-related careers into the classroom to share mathematics that may not typically be covered in the K–12 curriculum. In the fall of 2009, a colleague of mine, Dr. Rishi Nath, and I did just that at the college level. We started a mathematics circle called the York Tensor Scholars at York College, City University of New York (CUNY). It had two goals: (1) introducing varied mathematical topics to challenge the traditional views of mathematics and how it is defined and (2) challenging perceptions of who can and does excel in mathematics. We invited speakers, almost exclusively women, whose work focused on a broad range of mathematics and covered areas not typically introduced in K–12 curriculum or undergraduate courses for math majors. As an example, Dr. Amanda Redlich, then a postdoctoral researcher at Rutgers University and now an assistant professor at the University of Massachusetts, gave a talk about how knitting can be used to model three-dimensional surfaces. In her talk, we learned about the Kitchener stitch, which in knitting is used to join (graft) knitted pieces together seamlessly, as in a sock with a seamless toe. Dr. Redlich uses the stitch and knitting in general to create models of surfaces with various properties. Dr. Diana Thomas, a professor of mathematical sciences at the U.S. Military Academy, gave a talk about her creation of mathematical models to study changes in body composition during weight change. We also invited Dr. Vrunda Prabhu, then a professor at Bronx Community College, to give a talk about mathematics and creativity. These examples feature mathematics that is being done in the present day but that, typically, does not enter the discussion of the subject at the school or collegiate level.

Additionally, there were talks on graph theory, number theory, cryptography (making and breaking codes), the mathematics of Wall Street, and the mathematics of the steel drum, and even one where the researcher applied the way lightning bugs align their blinking to the synchronization of clocks. The topics were incredibly diverse. At the program's conclusion, I conducted a study of the program that included surveys of those who had attended our events as well as interviews of some of the students who participated in the program. Student survey responses confirmed our belief that students would find these topics both interesting and new. In fact, students often remarked that the mathematics they were being exposed to was different from that to which they were accustomed in their coursework.

However, despite exposure to these very different topics in mathematics, at the conclusion of their time in the program, those students who were interviewed still clung to very traditional beliefs about what mathematics is. When asked to define it, most students linked mathematics to numbers, equations, and formulas, as evidenced by responses such as the following:

- *Numbers. What else is math? It's the study of numbers and how they relate to each other.*

- *It's a bunch of numbers and formulas put together.*

Of the ten students interviewed, there were only two whose answers showed a less narrow focus. One student described mathematics as "the foundation" and when pressed added that "mathematics is the foundation of everything." Another student described a change in their view of mathematics. This student explained that prior to participation in the group, they held a view of mathematics as consisting of equations to be solved, but "Now I'd say that mathematics is the study of patterns." It was this type of progression that I had hoped to see in the students who participated in the program, but it manifested itself in only one of the students interviewed. Why might that be?

Given that the featured topics challenged the traditional perceptions of mathematics and that the students lacked familiarity with such topics, one might expect that students' own views of mathematics would be broadened. Yet, in terms of beliefs about mathematics, most students still clung to the traditional definition. This is a very common outcome but not one we expected from a group of students who over the course of several years had been exposed to much more than just numbers and equations through the work of mathematicians currently in the field. However, it seems that years of schooling that prioritized numbers and algebra could not be undone in a program where students spent an average of two years. We must start challenging traditional conceptions of mathematics and building a broader set of mathematical experiences for students from a *much* earlier age.

To challenge traditional views of mathematics in the classroom, start a conversation with students about what they perceive math to be and what math really can be.

Try using the following prompts to get students talking about mathematics and their experiences with it.

- The best thing about math is . . .
- The worst thing about math is . . .
- Learning math is like . . .

Consider the responses you obtain and how you might work with them. Do they point to negative or stereotypical views of mathematics? If so, consider challenging these perceptions by using more diverse problems or by bringing in individuals who use mathematics in their lives to share that mathematics with students. Do they point to stereotypical views of mathematicians? If so, consider bringing in the stories of diverse mathematicians from underrepresented groups through readings, posters, and films. Do they point to negative views around the learning of mathematics? If so, consider activities that encourage productive struggle and attempt to build student confidence in addition to mathematical competence. Are student answers full of creativity and positive views of mathematics, mathematicians, and the students themselves as doers of mathematics? Some surely will be. Build on these by highlighting them often. Use students' own words to remind them of the positive when it comes to mathematics. Hang their words on your walls and reference them frequently.

THE FOCUS OF K–12 MATHEMATICS AND WHAT GETS LEFT OUT

There are consequences to the narrowing of mathematics to numbers and equations. One of these is the fact that students leave their formal schooling with an incomplete view of what mathematics is and of the reality that it is constantly growing and evolving. This may mean that individuals who otherwise would have been attracted to the field, or who would have seen the field as a living one, may be put off by the focus on numbers and equations and thus shy away from it. This may result in individuals not engaging with it beyond what is required and even shying away from the topic as adults, though there exists an abundance of what are called *popular mathematics* books written for general audiences with no presumption of mathematical background that bring the subject to light and show the connections between mathematics and other fields in exciting ways. Lack of exposure during K–12 schooling to perfectly valid, beautiful, and in many cases useful mathematics that lies somewhat outside the field of algebra may lead many to think of this vibrant field as dull and cold. It might also lead some to believe they are *bad at math* when in reality they may struggle in an area commonly taught in school but excel in other areas that are often given less time or left out completely.

Resources: Popular Books That Connect Math to the Real World

Ellenberg, J. (2015). *How not to be wrong: The power of mathematical thinking.* Penguin.

Fry, H. (2015). *The mathematics of love: Patterns, proofs, and the search for the ultimate equation.* Simon & Schuster.

Goodreads. (2022). *Popular mathematics books.* https://www.goodreads.com/shelf/show/popular-mathematics

Parker, M. (2021). *Humble pi: When math goes wrong in the real world.* Penguin.

Suzuki, J. (2015). *Constitutional calculus: The math of justice and the myth of common sense.* JHU Press.

As an example of the curricular focus I am talking about, at the end of kindergarten my daughter came home with several workbooks that she had used during the year in her class. One of these was a math workbook. I noticed that in most of the workbooks there were some pages that had not been completed, and I decided we would complete them during the summer. Upon opening her mathematics book, I was pleased to see that almost every topic had been covered. Actually, only one had not been touched. All the content that had to do with numbers (counting, place value, decomposition, relative size, etc.) and with equations (numeric equations involving addition and subtraction as well as word problems that could be solved using these equations) had been completed. The one topic that had not been addressed was geometry. Specifically, content focused on shapes and their properties along with three-dimensional solids and their properties was skipped. A review of mathematical workbooks geared to young children reveals that geometry and other topics such as time, money, and sequencing tend to be found at the back of the book while content related to numbers and equations comes up front. Even in the kindergarten curriculum, long before students see advanced mathematics, the bias toward numbers and equations is clear. A review of the Common Core State Standards for Mathematics (Common Core State Standards Initiative, 2021) reveals the majority of standards in Grades K–8 focus on number and operations, operations and algebraic thinking, and expressions and equations. There is much less of an emphasis on geometry, probability, and statistics and virtually no topics in discrete mathematics, a rich area in the field that we will revisit later in this chapter. Likewise in Canada, a comparison of the Common Core State Standards to the Ontario content expectations, the Québec Essential Knowledges, and the Western and Northern Canadian Protocol curriculums conducted by the National Council of Teachers of Mathematics (2013) show, similarly, a focus on algebra, number sense, and numeration while showing a lack of discrete mathematics topics. If anything, less geometry is covered in the Canadian standards compared to the American ones, though there seems to be more of a focus on measurement. It is interesting to note that in 2020 the mathematics standards adopted in Ontario included financial literacy in the curriculum for Grades 1–8 (Ontario Ministry of Education, 2021). Specifically, it was stated that students would understand the value of money over time, learn financial well-being, and work on creating and managing budgets. At the high school level, the focus in schools across the United States and Canada is on algebra, although there is more geometry than in the younger grades. There are virtually no topics in discrete mathematics—including in election theory, which feels necessary for students in a democratic civil society. Similarly, there is little, if any, financial mathematics. Probability and statistics, though included in the Common Core State Standards and in the various standards documents across Canada's provinces, do not get a deep treatment.

Making matters worse is the fact that in many areas, fewer than four years of math are required at the high school level. Canadian requirements vary by province, with most requiring three or four years of mathematics. In the United States, most states require three years, but some, such as California, require only two. What this typically means is that students who struggle the most tend to not take mathematics beyond the required number of years (except to repeat a course they did not pass), which means that those needing the most help in the subject take the least mathematics courses and are often not exposed to the subject in the year preceding college, making it difficult for them to be successful in their entry-level mathematics course. In low-performing schools, it is common for an algebra course to be spread out over two years and for the two years to be coded differently so that students get credit for two years of mathematics even though they are being exposed to what is typically taught in a one-year course. Even after taking geometry in their third year, these students have completed the course requirement for mathematics but have studied the equivalent of only two years of high school mathematics, impacting their ability to be admitted to and do well in college. With a focus on algebra in preparation for calculus, many students leave high school with only a surface-level understanding of other branches of mathematics.

> With a focus on algebra in preparation for calculus, many students leave high school with only a surface-level understanding of other branches of mathematics.

These include discrete mathematics, financial mathematics, and statistics. Additionally, data science, which aims to provide insight from large data sets that are increasingly prevalent in our social world, is increasing in importance and has been proposed as part of California's K–12 math framework. These areas contain valuable mathematical content that can be applied across numerous topics.

The Value of Discrete Mathematics

As an example, in discrete mathematics students study graph theory. A graph is a collection of vertices, some of which are connected to others by edges (sometimes directed and sometimes without a direction). One example of this is a subway map: The vertices are the subway stations where the trains stop; trains that you can take from one station to the next are connected by an edge; some stations are hubs that connect multiple train lines. These graphs can be used to visually describe the flight paths of airplanes, the connections between major freeways, and more, facilitating efficient travel and ensuring that both individuals and goods can make it quickly and safely to the desired location. For example, see the Canadian rail map in Figure 2.2.

Figure 2.2 • *Canadian Rail Map*

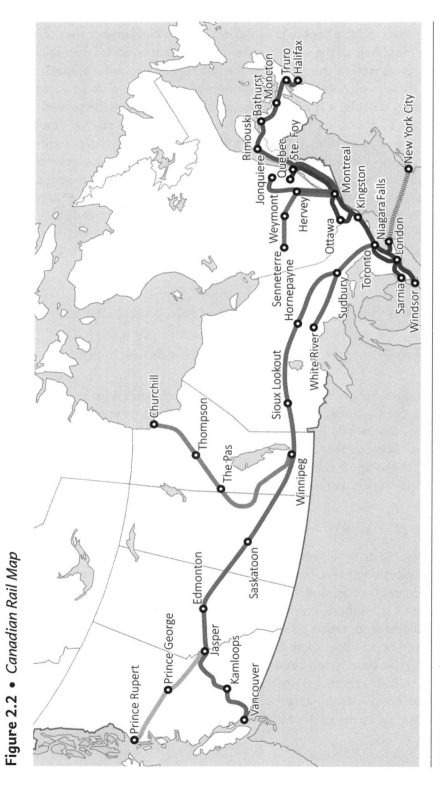

SOURCE: Stepney, C. (2019, February 1). VIA network map 2019. In *Wikipedia*. https://en.m.wikipedia.org/wiki/File:VIANetworkMap2019.png; CC by 4.0 (https://creativecommons.org/licenses/by/4.0/)

Similarly, you can create a graph of this type to model the spread of disease. If person A gives the disease to person B, there is a directed edge from A to B. Thus, graphs can be used for contact tracing such as was implemented in response to the COVID-19 pandemic. Directed graphs can also be used to model social networks. Individuals with many followers have many edges leading into the vertex, while those who follow many people have many edges leading out.

Discrete mathematics also includes election theory. Dr. Joseph Malkevitch (2008), professor emeritus at York College, CUNY, showed incredibly that using the same set of ranked ballots (where each voter is asked to rank the candidates in order from their favorite to their least favorite) could produce five different winners in a five-person race depending on the method used to evaluate those ballots. In the United States, it is often the case that the candidate with the most votes wins (plurality voting), but there are cases, such as the Electoral College, where this is not true. Every political party in Canada uses ranked voting to elect its leaders. Here people indicate their first, second, and third choice. Initially, all the first choices are added, and if there is a majority, that person wins. If not, the candidate with the fewest votes is eliminated, and those ballots where that person was ranked first now get transferred to whomever was ranked second. The process repeats until someone has a majority. There are also races where the candidate with the most votes does not necessarily win certain elections unless they have more than a certain percentage of the vote. Thus, in a race with many candidates, there may need to be a runoff election where a subset of the top candidates face off against each other. In the United States, the election of a president relies on the use of the Electoral College. Students would benefit from learning about this, especially given the fact that many if not most will be expected to participate in the political life of their country as voters themselves.

Resources: Discrete Mathematics

DeBellis, V. A., Rosenstein, J. G., Hart, E. W., Kenney, M. J., & House, P. A. (2009). *Navigating through discrete mathematics in prekindergarten–Grade 5: Principles and standards for school mathematics navigations*. National Council of Teachers of Mathematics.

Hart, E. W., Kenney, M. J., DeBellis, V. A., & Rosenstein, J. G. (2008). *Navigating through discrete mathematics in Grades 6–12: Principles and standards for school mathematics navigations*. National Council of Teachers of Mathematics.

THE INCREASING VALUE OF STATISTICS

The proliferation of numeric data in every aspect of social life places an additional burden on citizens with respect to voting. Two candidates with very different plans when it comes to taxes may both argue that their plan lowers taxes for the middle class while generating sufficient revenue to sustain these cuts over several years. Ascertaining the validity of each claim, as well as understanding the impact of such policies on one's own circumstances and on society, requires a good deal of mathematical fluency. Conflicting claims from opposing candidates are commonplace, and being able to examine these critically is essential to being able to vote for the candidate who most aligns with one's own views of major issues.

> It is increasingly the case that candidacies in the modern era can be won or lost based on the unemployment rate, the crime rate, or the Dow Jones index. Our multitudes of numerical indicators summarize the complex economic, political, and social health of the country, and citizens need to be able to decode and decipher this modern day "political arithmetic." (Cohen, 2003, p. 7)

In addition, increasingly complex visual displays of data are being incorporated into newspapers, magazines, and other media outlets to shed light on our political systems, yet without the ability to properly make sense of them, such information becomes confusing rather than clarifying.

There is currently a move toward offering both Advanced Placement (AP) Calculus (what had been the traditional offering) and AP Statistics in schools. While this trend is a welcomed one, the only students benefiting are those who have access to AP courses, and there is an underrepresentation of Black and Latinx students in these courses. As such, Black and Latinx students are less likely to be exposed to the rigorous study of statistics. Additionally, AP Statistics has traditionally been taught as a theoretical course, with little direct intentional applications to current societal realities. While it may be a good theoretical course, it has not traditionally been taught in ways that highlight how statistics can be used to understand, analyze, and explore the social realities of the current day.

Resources: Statistics and Data Science

American Statistical Association & National Council of Teachers of Mathematics. (2022). Statistics Teacher *publications*. https://www .statisticsteacher.org/statistics-teacher-publications/

This Is Statistics. (2022). *Statistics resources for educators*. American Statistical Association. https://thisisstatistics.org/educators/

The Value of Financial Mathematics

Another beneficial area of mathematics that is often left out of the curriculum is financial mathematics or, as some call it, financial literacy. While it may seem that financial mathematics is basic and that students can pick it up as they go, this is not necessarily the case. Compound interest, annuities, stocks, and paying off one's credit card rely on mathematics that includes geometric sequences, exponential and logarithmic functions, and even calculus. As such, explicit instruction in financial mathematics would benefit students. Individuals from wealthier backgrounds are more likely to invest in the stock market and have a retirement plan, as well as to have set up other methods of both savings and generating income from those savings. Their children, through lived experience, are being taught the value of these investments. Individuals from poor and working-class backgrounds are not nearly as often being taught these lessons from lived experiences. Further, they are more likely to be in debt. Granted, teaching financial math won't necessarily get folks out of debt. There exists a myriad of reasons why wealth is concentrated among so few. We certainly need to think through the policies that created the wealth disparity we see in our society and institute social policies that promote fair wages and ensure jobs with good benefits while reconsidering our tax structure so that everyone pays their fair share. However, having a solid understanding of how wealth works and the mathematics behind various financial realities can help individuals to make better financial decisions as well as advocate for policies and programs from which they may benefit.

Resources: Financial Mathematics

Andal, W. (2021). *Finance 102 for kids: Practical money lessons children cannot afford to miss*. Gatekeeper Press.

FINRA Investor Education Foundation. (2022). *Resources for educators*. https://www.finrafoundation.org/people-we-help/resources-for-educators

Stephenson, A. (with Mills, L.). (2020). *Teach your child about money through play: 110+ games/activities, tips, and resources to teach kids financial literacy at an early age*. SimplyOutrageousYouth.org

BALANCING ALGEBRA AND OTHER AREAS OF MATH

At present, in a society that values algebra so heavily and relies on it for graduation requirements, entrance into college, and beyond, not teaching students algebra puts them at a disadvantage. In fact, given that algebra is a predictor of future success, Moses and Cobb (2001) consider the access to high-quality upper-level mathematics, including algebra, for Black students as important as Black people earning the right to vote during the civil rights movement. They see it as a civil right. Further, that we should be taught only what we will use—the principle of utilitarianism in education—is difficult to support. We are never truly sure of what we will need to know going forward.

> Further, that we should be taught only what we will use—the principle of utilitarianism in education— is difficult to support. We are never truly sure of what we will need to know going forward.

More than this, however, is the idea of well-rounded individuals whose broad understanding of a wide range of subjects allows them to appreciate and readily engage with the world around them.

There is value and merit to the idea that we can reduce the focus on algebra to expand students' knowledge of mathematics be it quantitative literacy, financial mathematics, discrete mathematics, statistics, data science, or other areas whose study might broaden students' views of what mathematics is. This may also introduce students to topics that they would not otherwise see while highlighting the ways that mathematics connects to their lives. With this connection clear in their minds, it might be harder for individuals to buy into the *bad at math* trope.

When we move the focus, even just slightly, away from algebra, we find there is room in the curriculum for other perfectly valid and useful mathematics that can benefit students as they find their way in our social world. We should find ways to bring some of these areas into the classroom while pushing for changes to state-mandated curricula that value algebra and number sense over all other fields. There is room for a more inclusive treatment of all mathematics has to offer. Expanding what most individuals see as mathematics by ensuring that students are exposed to more varied mathematical topics might lead more people to realize that they do enjoy doing mathematics. What they think of as mathematics would broaden to include topics that might attract them.

Rethinking Calculus

The curriculum, as it stands, focuses on algebra with the eventual goal of calculus. AP Calculus is one way students position themselves for entry into college and obtain college credit. A score of 3 on the exam is usually accepted for college credit. However, access to these classes is limited for students from marginalized communities. Specifically, Black students represent just 8.8% of exam takers and 12.3% of those who scored a 3 or higher (Jaschick, 2019). Indigenous students account for less than 1% of exam takers and less than 1% of those scoring a 3 or above. The numbers are more encouraging for Latinx students (25.5% take the exam with 23.6% earning a 3 or above). If the eventual goal of mathematics is proficiency in calculus, then our most vulnerable students are being left out. This makes it difficult for them to succeed in science, technology, engineering, and mathematics (STEM) fields going forward. Think of the mathematical advances we miss out on as a society because some students are not given the opportunity to take courses that position them to pursue STEM fields. It isn't just the students themselves who miss out; it is all of us.

Recently, there has been some movement away from calculus as the solitary end goal of study in mathematics, and efforts at de-tracking mathematics education are underway. As an example, with the shift to the Common Core State Standards for Mathematics, and the resulting shifts in rigor and topics in the eighth-grade standards in particular, the San Francisco Unified School District eliminated accelerated mathematics courses in middle and high school mathematics. All students take a common math sequence of heterogeneously grouped classes in middle school and the first two years of high school. No longer is algebra offered in eighth grade (which positioned only those who had access to it more favorably in future studies), but rather all students take it in ninth grade and geometry in 10th grade. Going forward in their studies beyond algebra, students choose math courses based on their interests, including a broader array of advanced courses. Shifts in mathematics pathways at the high school level are also underway as part of the Catalyzing Change initiative of the National Council of Teachers of Mathematics. In this way, multiple paths toward the completion of mathematics requirements are being developed, affording students more choice and creating a more equitable system. No longer will calculus be the only way through.

At the college level, calculus is a course that students typically struggle with and one that has been used to weed out students from STEM programs. Seeing this as a problem, faculty in the life sciences at the University of California, Los Angeles, developed a two-semester calculus sequence that covers the typical content but does so while emphasizing its connection to biology and physiology (Burdman, 2022). Students who took the course received better grades in subsequent science courses than those enrolled in the traditional calculus sequence, and student interest in the content doubled. In a similar example at Ohio's Wright State University, traditional prerequisites for calculus were replaced with a contextualized math class that focused on real-world problems in engineering (Burdman, 2022). Here it was found that the course was particularly beneficial to students from historically marginalized communities in mathematics including Black and Latinx students. It might be time that we think about how to bring contexts into mathematics teaching that increase student interest while also making clear how the mathematics connects to our social world.

WHAT CAN YOU DO TO CHALLENGE THE BELIEF THAT MATHEMATICS IS JUST NUMBERS AND EQUATIONS?

A common refrain from educators is that the curriculum they are expected to teach is already so packed, there is no time for other topics. But these topics can be integrated in small ways and in larger ones. Consider the following:

- Ask students to solve some problems using various methods—geometrically, acting it out, describing it in words, using pictures, and so on—not just algebraically or numerically.
- Use problems from diverse areas of math.
- Fill your classroom with representations of diverse mathematics (see more in Chapter 3).
- Incorporate a problem of the week drawing from diverse areas in mathematics.
- Incorporate games that focus on geometry, spatial reasoning, logic puzzles, and so on (see mathforlove.com for some examples).
- Consider offering courses or additional emphasis in statistics, financial math, and discrete math at your school.
- Consider reordering your curriculum to expose students to geometry, probability, and statistics earlier in the year.

- Listen to podcasts that discuss a broader view of mathematics education, including *Make Math Moments That Matter* (Pearce & Orr, 2018–present), *Kids Math Talk* (Harrison, 2020–present), *Sum of It All* (Mendivil & Alcorn, 2021–present), and *DebateMath* (Luzniak & Baier, 2022).

- Check out and create your own events like Math On-a-Stick (https://talkingmathwithkids.com/mathonastick) or participate in Global Math Week from the Global Math Project (globalmathproject.org).

- Advocate for local- and national-level changes to the curriculum by getting involved in various political and educational organizations, such as your state affiliate of the National Council of Teachers of Mathematics (NCTM) or NCSM, Leadership in Mathematics Education.

Questions for Reflection

For Teachers

- How might you utilize problems and content from a wide range of areas in your classes?

- Can you reconsider how you organize the topics you teach to move up content that goes beyond algebra and number sense?

- How can you engage students in the study of discrete math, financial math, and statistics?

- What opportunities do your students have to see mathematics as more than numbers and equations?

- How can you bring popular texts in mathematics to your teaching?

For Instructional Leaders

- How can you work with teachers to help them balance algebra and number sense with other branches of mathematics?

- How can you work with teachers around leveraging real-world topics with statistics and election theory?

- What resources can you provide that challenge traditional views of mathematics?

(Continued)

(Continued)

For Administrators

- How can you support teachers who want to bring in areas of mathematics that are outside algebra and number sense?

- How might you create and/or strengthen courses, activities, events, and opportunities around financial mathematics, discrete mathematics, and statistics?

- In what ways can you ensure that the educational resources and curriculum adopted by your school reflect the diversity of the field of mathematics?

- What mathematical content is reflected in the professional development opportunities offered to the teachers at your school?

CHAPTER 3

MATHEMATICIANS AND MATHEMATICIANS IN TRAINING

In this chapter we will:

- Discuss and challenge traditional beliefs about who mathematicians are.

- Consider representations of mathematicians in the media.

- Examine degree attainment in mathematics among underrepresented groups.

- Reflect on how you can challenge traditional beliefs about mathematicians and support those typically marginalized in mathematics.

HOW DOES SOCIETY VIEW MATHEMATICIANS?

With increased visibility of the stories of mathematicians who are women, Black and Latinx conceptions of mathematicians are slowly changing. Still, with respect to who can and does excel in mathematics, there are very traditional views of mathematicians. We have, as a society, created a characterization of mathematicians that involves being

- unusually smart,

- able to recognize or see patterns that others do not, and

- able to compute in one's mind with both precision and speed.

Further, mathematicians are often portrayed as socially awkward individuals who, despite their brilliance, don't quite fit in or know how to navigate the social norms *normal* folks do. They are also, typically, portrayed as white males. In their study, Helen Forgasz and Gilah Leder (2017) asked people who they thought were better at mathematics: girls or boys. Across three countries (Australia, Canada, and the United Kingdom), the percentage of people saying girls and boys were equal in ability ranged from 31% to 45%, and in all cases among those who did see a perceived difference, boys were thought to be more able. Among Canadians specifically, the percentage of those who thought boys were better than girls at math was roughly 30%.

In the book *Inventing the Mathematician: Gender, Race, and Our Cultural Understanding of Mathematics*, Sara Hottinger (2016) explores characterizations of mathematicians across several spaces. These include math textbooks, as well as texts about the history of mathematics, portraiture, and ethnomathematics. Hottinger explains that as a society, Western culture is based on the ideal of the rational, mathematical mind and that this is inextricably linked to males, most especially white males. She discusses the books of Danica McKellar, which aim to bring mathematics content to girls in a way that is engaging and intertwined with their interests. Specifically, these are *Math Doesn't Suck: How to Survive Middle School Math Without Losing Your Mind or Breaking a Nail* (McKellar, 2007), *Kiss My Math: Showing Pre-algebra Who's Boss* (McKellar, 2009), *Hot X: Algebra Exposed* (McKellar, 2010), and *Girls Get Curves: Geometry Takes Shape* (McKellar, 2012). The books are flawed in that they narrowly value and portray the experiences of girls who are white and middle-class with stereotypical interests in boys and fashion, but they are at least an attempt at providing one alternative narrative with respect to who can do mathematics. However, we need to go much further. Those at the intersection of marginalized identities—as an example, Black girls—are left even further out of the conversation. It isn't just that we are stereotyping women, but the erasure of girls of color from our attempts at changing the narrative of who can do math does little to disrupt the stereotypes around race that exist in our society. They lead to further marginalization that affects not only the students themselves but all of us. We lose the potentially valuable contributions that these individuals might make to move mathematics and society forward in positive ways.

The *white male math myth*—as David Stinson (2013) refers to it—persists. If you do an online search for images related to the word *mathematician*, you will see an overwhelming majority of images of white men. There are, of course, numerous examples of mathematicians who defy the traditional stereotypes, and there have been for centuries. However, for years their names and stories have not been made widely known.

> There are, of course, numerous examples of
> mathematicians who defy the traditional stereotypes, and
> there have been for centuries. However, for years their
> names and stories have not been made widely known.

One notable exception is the story of Katherine Johnson, Mary Jackson, and Dorothy Vaughan, who worked for the National Aeronautics and Space Administration (NASA) in the 1950s and 1960s and whose work played a pivotal role in making space travel possible. Their story is the focus of the book *Hidden Figures: The American Dream and the Untold Story of the Black Women Who Helped Win the Space Race* (Shetterly, 2016), which was turned into the film *Hidden Figures* (Melfi, 2016). Similarly, a children's book, *Maryam's Magic: The Story of Mathematician Maryam Mirzakhani* (Reid, 2021), tells the story of Maryam Mirzakhani. In 2014, as a professor of mathematics at Stanford University, she made history as the first woman and the first Iranian to receive the Fields Medal—mathematics' highest prize, which is often referred to as the Nobel Prize of mathematics (there is no actual Nobel Prize in mathematics). In 2022, Maryna Sergiivna Viazovska, a Ukrainian mathematician known for her work in sphere packing, became the second woman to receive the prize. It is my hope that in the coming years, more stories of women in mathematics come to light and more women and people of color are recognized for their contributions to the field.

REPRESENTATIONS OF MATHEMATICALLY ABLE CHARACTERS IN THE MEDIA

On television, we still mostly see mathematically able characters portrayed as white males who are socially awkward geniuses gifted with natural talent and who are unlike the rest of us. As an example, the television show *Scorpion* (Woodrow et al., 2014–2018) features a cast of geniuses hired by the U.S. Department of Homeland Security to solve a series of complex problems. Among them are a mathematician, a computer scientist, an engineer, and a psychologist. They are assisted by a non-genius former waitress who helps them with the social aspects of life and who herself has a young child who is also a genius. The main character, Walter O'Brien, is a white male with an unusually high IQ who as a child hacked into NASA's mainframe in search of blueprints to hang on his wall. He is a computer genius who is socially awkward and has a very low emotional quotient, often having difficulty in social situations and being unable to both express his emotions and understand the emotions of those around him. Another incredibly bright character on that show is the mathematician Sylvester Dodd, who struggles with obsessive-compulsive disorder, anxiety, and a fear of germs.

Another show that includes mathematically able characters is *The Big Bang Theory* (Lorre et al., 2007–2019) about a group of scientists who are friends. Here, too, we see a very traditional representation of scientists. Of the five main characters who are scientists, four are white (one is Indian), and four are men. All are portrayed as nerdy, not quite like the average person, and all struggle with social norms in some way. Sheldon Cooper, a theoretical physicist, was a child prodigy with an IQ of 187 who began college at the age of 11 and completed his PhD at 16, yet he finds many social situations difficult to comprehend and has a ritualized way of living. For example, he always sits in the same spot on the sofa and has great difficulty when someone takes his spot, eventually forcing them to move because he cannot bring himself to sit elsewhere. The only scientist on the show who is a woman displays little emotion and is awkward in social situations, especially when first introduced. That these characters display some characteristics of neurodivergent individuals is troubling as well because it feeds into stereotypes around mathematically able people as well as stereotypes of neurodivergent people—namely, that mathematically able people tend to be neurodivergent and that those who are neurodivergent excel in math and science. This fails to capture the nuances and diversity of both populations.

On television recently there have been several more progressive characters. For example, Patterson on *Blindspot* (Schecter et al., 2015–2020) is a white woman who uses her computing and mathematical skills to decode clues hidden in the tattoos of a woman that direct the Federal Bureau of Investigation (FBI) to cases they must solve. Penelope Garcia on *Criminal Minds* (Gordon et al., 2005–2020) is an incredibly skilled computer hacker who gets recruited by the FBI. As well as being a very emotional and quirky Latina, she uses her skills to narrow the pool of suspects the team she works with is investigating. Despite these examples, however, there continues to be a need for more diverse mathematically able characters on television. As far as I can tell, there is no Black, Latina, or Indigenous woman as a main character on a television series to date who is depicted as a successful mathematician. There is, however, one in film. Shuri, from the film *Black Panther* (Coogler, 2018), is T'Challa's (Black Panther's) younger sister, a brilliant scientist who develops technology for the advanced civilization of Wakanda and is a Black woman.

REPRESENTATIONS OF SCIENTIFICALLY ABLE CHARACTERS IN CHILDREN'S MEDIA

It is interesting, and somewhat hopeful, to note that there is some representation of nonwhite scientifically minded girls in children's media, but here, too, this is the exception and not the rule. The Disney princess Elena, from the show *Elena of Avalor* (Gerber & Mitchell, 2016–2020), is described as a Latina princess. Her sister, Isabella, is an inventor. In the Barbie® television series, *Barbie Dreamhouse Adventures* (Keenan et al., 2018–present), Barbie's friend Teresa is very much into science and often provides scientific insights for her friends. Teresa is Latina and, at times, speaks Spanish on the show. Also of note is the Canadian TV show *The Odd Squad* (Johnson et al., 2014–present), which features a group of kids who work in an organization based in Toronto that investigates and fixes anything odd that happens in the world. A comedy, it aims to introduce children to mathematical concepts like fractions, percentages, and various arithmetic functions. The *odd* in the odd squad refers to the problems themselves and not the mathematically minded children who explore them. Media representation of mathematically able characters is important as it affects mainstream conceptions of who can and does excel in the subject.

These misconceptions are evident outside the media as well. In 2013, The Children's Place featured a T-shirt marketed to girls with a checklist of items under the heading "my best subjects." There were shopping, music, dancing, and math. Of these, all but math was checked, and the phrase "well, nobody's perfect" was written underneath. That girls are expected to be good at shopping, music, and dancing but not math was troubling to many, and the shirt was eventually pulled from the shelves. In 2011, a shirt with the phrase "too pretty to do homework" was pulled by JCPenney for similar reasons. That these characterizations are a consistent presence in our children's lives can serve to drastically limit our potential for considering other narratives in terms of brilliance and accomplishment in the field.

Resources: Children's Books Featuring Women Mathematicians

Bardoe, C., & McClintock, B. (2018). *Nothing stopped Sophie: The story of unshakable mathematician Sophie Germain*. Little, Brown Books for Young Readers.

Becker, H., & Rust, K. (2020). *Emmy Noether: The most important mathematician you've never heard of*. Kids Can Press.

(Continued)

(Continued)

Jones, S. M. (2019). *Women who count: Honoring African American women mathematicians.* American Mathematical Society.

Mosca, J. F. (2019). *The girl with a mind for math: The story of Raye Montague.* Scholastic.

Reid, M. (2021). *Maryam's magic: The story of mathematician Maryam Mirzakhani.* HarperCollins.

Shetterly, M. L. (with Conkling, W.), & Freeman, L. (Illustrator). (2018). *Hidden figures: The true story of four Black women and the space race.* HarperCollins.

HOW ARE MATHEMATICIANS REPRESENTED IN TEXTBOOKS?

Hottinger (2016) also analyzes how female characters are depicted in mathematics textbooks for elementary and middle school students by studying several of the most popular texts used. She notes that the male characters are often presenting a new and novel way of arriving at solutions to a problem, while the female characters are often used in problems where the student is asked to repeat an existing procedure. Further, they are often characterized as helpers. From the time we are children we are bombarded with racist, gendered, and ableist stereotypes of who can and does excel in mathematics. It is no surprise, then, that many students often do not see themselves as doers of mathematics and avoid study in mathematics as well as mathematics-related careers.

> From the time we are children we are bombarded with racist, gendered, and ableist stereotypes of who can and does excel in mathematics. It is no surprise, then, that many students often do not see themselves as doers of mathematics and avoid study in mathematics as well as mathematics-related careers.

As an exercise, take the time to look at the representations of mathematically able people in your classroom. Do the posters on your walls, the characters in the books in your library, and the textbooks you use represent a wide variety of diverse mathematically able individuals? Is there evidence of stereotypical portrayals of mathematically able people, and if so, how can you work to change this?

WORKING TO SHIFT STUDENTS' VIEWS

In Chapter 2, I introduced the York Tensor Scholars program. Another aim of the program was to challenge the underrepresentation of women in mathematics and to push back against commonly held assumptions of who can and does do mathematics. To do this, the speakers we invited were almost exclusively women in mathematics who were welcoming and open to sharing their work and their experiences with our students, a large proportion of whom identified as Black, Latinx, and/or women (see Figure 3.1).

Figure 3.1 • *Percentage of York College Students Who Identify as Women, Black, or Latinx*

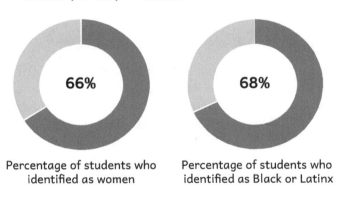

Percentage of students who identified as women

Percentage of students who identified as Black or Latinx

As these are all underrepresented groups in mathematics, the college was an ideal setting for such a program. The program was open to all interested students, and students did not have to be part of any underrepresented group in mathematics or be math majors to take part, but due to self-selection, many were. In fact, of the 51 individuals who participated in the program, only one was a white male, the traditionally represented group in the mathematical sciences, and 80% of the students were majoring in science, technology, engineering, and mathematics (STEM) fields (see Figure 3.2).

A subgroup of students in the program participated in interviews focused on how their experiences with the York Tensor Scholars informed their views of mathematics, mathematicians, and themselves as doers of mathematics.

Figure 3.2 • *Percentage of York Tensor Scholars Who Identify as Women, Black, or Latinx*

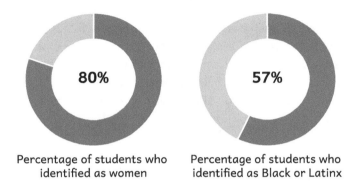

Percentage of students who identified as women   Percentage of students who identified as Black or Latinx

Students' Changing Views of Mathematicians

Most students expressed a change in their perception of mathematicians as a result of their participation in the group. Where initially students held to traditional views of mathematicians as male, older, graying, and socially awkward geniuses, their ongoing interactions with mathematicians especially at social outings led to a view of mathematicians as potentially funny, engaging, younger, and willing to spend time with students such as themselves. One student commented that she "did not expect to see a person with a good sense of humor," and another remarked that she did not expect the mathematicians to be "very social and open to questions." These responses are representative of those of the vast majority of students interviewed and show that our choice of speakers challenged the traditional view of mathematicians and further led to students actively challenging their own views of mathematicians. The change in beliefs about mathematicians toward people who are more normal—more like them—was apparent in many of the interviews as is highlighted in the following exchange.

LG: Did your opinion about mathematicians change in any way as part of your involvement in the Tensor group?

Student: Yeah.

LG: The people themselves?

Student: Well, the—yes. They are regular people, you know.

LG: Did you think they were regular people—before?

Student: No . . . Well, before I thought mathematicians were weird, you know, like they're always sitting, reading, you know. Doing nothing, just studying. But now I—I still think they are weird, but they are normal, too. You know? They can have fun. They can do other things. And they are normal, normal people, you know. I don't think they are just isolated like I thought before. I thought they were people who liked to be by themselves, you know, not approachable. I was . . . scared of them.

As you can see, this student originally characterized mathematicians as weird, isolated, and engaged in nothing but study, but in time this changed. This shift in perception was evident in all the students interviewed. Even where students' perceptions of what mathematics is proved harder to shift, as described in Chapter 2 participation in the group led students to reimagine mathematicians as normal people who are approachable and relatable.

Students' Views of Themselves as Mathematicians in Training

Irrespective of their major, many students indicated that they valued the ability to socialize with peers and others who share a common interest in mathematics, and that participation in the group afforded them this opportunity. While it might seem that this social connection alongside shifting one's views of mathematicians might translate into students seeing themselves more positively as doers of mathematics, sadly this was not the case. They did not refer to themselves as mathematicians, and all but one did not define themselves as mathematicians in training. When characterized this way by others, the label felt uncomfortable. That feeling of belongingness that they felt within our small community did not translate into a feeling of belongingness within the greater mathematical community.

A lot of people say that when I tell them I'm a math major. They're like, "So, you're going to be a mathematician." And it takes a second to take it in, like, "Yeah, I am." Like when I think about a mathematician, I think about Isaac Newton, Albert Einstein. Like I don't see myself as one of those. So, it's hard to believe it.

It is interesting to note that the individuals this student lists as mathematicians are Newton and Einstein, both of whom fit traditional characterizations of mathematicians. The student who made this statement is a young woman of Caribbean descent who planned to teach mathematics at the elementary school level. Though indicating that she is good at math, this student still distanced herself from those she considers mathematicians—especially those for whom, according to her, the material comes naturally.

This research highlights the reality that a year or two of participation in a program such as this at the collegiate level cannot necessarily undo years upon years of conditioning about who can and does excel at mathematics. For these students, it came too late. It is for this reason that we must expose students early to diverse, mathematically able people and to foster in them a sense of belongingness with respect to the discipline. Exposing students from a young age to diverse characters in books or people throughout history who are strong in mathematics and who resemble them in some way—be that with respect to gender, culture, race, ability, age, and so on—is invaluable. The earlier we do this, the better.

UNDERREPRESENTED GROUPS IN MATHEMATICS

People who identify as women, Black, and Latinx are underrepresented in mathematics. This is well documented. The National Science Foundation defines underrepresented groups as those whose representation in mathematics degree programs and math-related careers is significantly lower than their representation in the U.S. population (National Center for Science and Engineering Statistics, 2019). They include in this categorization women, persons with disabilities, and those from three racial and ethnic groups (Black, Latinx, and American Indian or Alaska Native). The percentage of doctoral degree recipients in science and engineering (which includes mathematics) awarded to members of underrepresented groups in U.S. schools is in the single digits, though these groups make up 33.9% of the population ages 18–64 in the United States (see Figure 3.3).

Let's take some time to explore that underrepresentation. Women from all backgrounds attend college and persist to degrees at higher levels than their male counterparts both in the United States and in Canada. The biggest difference is not in degrees attained but in the majors students choose to pursue. Still, with respect to mathematics and statistics, 42.4% of the bachelor's degrees and 41.6% of the master's degrees in the United States are earned by women (National Center for Science and Engineering Statistics, 2019). Here, however, the issue is a leaky pipeline where, though they have strong showings in bachelor's- and master's-level degrees, only 28.5% of the doctoral degrees in mathematics and statistics are earned by women. Next, let's consider the percentage of degrees earned by Black and Latinx women using Figures 3.4 and 3.5.

Figure 3.3 • *Doctoral Degrees in Science and Engineering Awarded to Those From Underrepresented Groups in 2019 From U.S. Institutions*

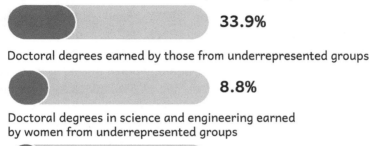

Population in the U.S. from underrepresented groups

33.9%

Doctoral degrees earned by those from underrepresented groups

8.8%

Doctoral degrees in science and engineering earned by women from underrepresented groups

5%

Figure 3.4 • *Degrees in Science and Engineering by Black Women From U.S. Institutions in 2019*

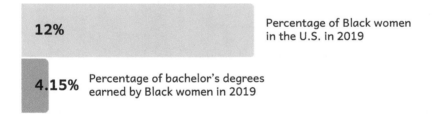

12% — Percentage of Black women in the U.S. in 2019

4.15% Percentage of bachelor's degrees earned by Black women in 2019

Figure 3.5 • *Degrees in Science and Engineering by Latinx Women From U.S. Institutions in 2019*

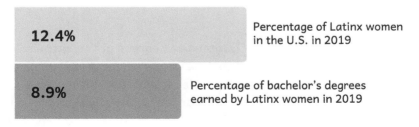

12.4% — Percentage of Latinx women in the U.S. in 2019

8.9% — Percentage of bachelor's degrees earned by Latinx women in 2019

That Black and Latinx women are underrepresented in mathematics is extremely clear. Further, as the percentages in these figures are for bachelor's degrees, one can safely assume that the percentage for doctoral degrees attained is less, a very troubling reality.

A study by Winnie Chan and colleagues (2021) relied on data from the 2011 high school graduation cohort to examine enrollment in and persistence to bachelor's degrees in STEM fields between men and women across Canada (see Figure 3.6). Here too, the percentage of women enrolling in university was higher than that for men (39.6% versus 31.5%), but differences exist in what programs students enroll in and graduate from. Women were 10 percentage points less likely to graduate with a degree in STEM than men.

Figure 3.6 • *Percentage of Bachelor's Degrees Awarded to STEM Students From Canadian High School Graduating Class of 2011*

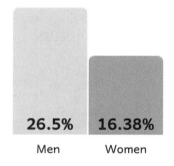

26.5% 16.38%

Men Women

The researchers established different definitions of STEM and found that the difference between men's and women's bachelor's degree attainment was even more pronounced if considering a math-intensive STEM field such as physics or mathematics. Men were over four times more likely to graduate with a bachelor's degree in a math-intensive field than women (see Figure 3.7).

Given the stark differences in math-intensive degrees at the bachelor's degree level and the leaky pipeline that exists, as we progress to master's degrees and doctoral degrees the gap widens. Data on STEM degree attainment by race in Canada are much harder to come by than in the United States, as there has not been a history of collecting data pertaining to race and educational attainment in Canada. This has many calling out for change.

Figure 3.7 • *Percentage of Bachelor's Degrees Awarded in Math-Intensive STEM Fields to Canadian Students From the High School Graduating Class of 2011*

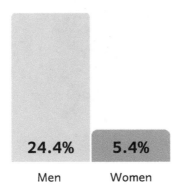

One such person is Jennifer D. Adams, Canada Research Chair in Creativity and STEM and an associate professor at the University of Calgary. She writes, "Although there's a history of research on under-representation in STEM the needle hasn't moved much. This is because existing research and policy has focused on assimilating students rather than examining existing societal and institutional barriers to STEM majors and careers" (Adams, 2021).

THE HARMFUL EFFECTS OF MARGINALIZING CERTAIN GROUPS

The numbers just discussed do not reflect the mathematical capabilities of those from underrepresented groups, but rather reflect the fact that these individuals have been marginalized, denied opportunities, and pushed out of the subject. To examine the harmful effects and lost opportunities of this marginalization, let's look at a couple of examples.

Sophie Germain and Eugenia Lofton Haynes

Historically, it was believed that studying mathematics was harmful to women, but there have been many women who tenaciously broke through (or went around) barriers and made significant advancements in the field. One such case is 18th-century mathematician Sophie Germain (see Figure 3.8). Germain, who was born in France in 1776, would sneak into her father's study to learn mathematics from his books.

Figure 3.8 • *Coin Featuring Sophie Germain*

SOURCE: Store Norske Leksikon

Fearing that this would be harmful to her, her parents took away her clothes and candles to make it more difficult for her. Despite this, she continued her studies, learning from numerous texts and eventually developing her own mathematical theorems. As a high school teacher, when I shared her story with students, I often asked if there was anything in their own lives they would be this willing to work hard to pursue. Their answers were often illuminating, and their connection to Germain's life grew as a result. Under an assumed name, Germain started correspondences with various mathematicians at the time including Joseph-Louis Lagrange, Adrien-Marie Legendre, and Carl Friedrich Gauss, with whom she shared some of the mathematics she was developing. She made contributions to elasticity theory, including a paper that earned her the grand prize from the French Academy of Sciences. She also made significant mathematical contributions to number theory and has a class of prime numbers that are named for her. A prime number p is a Germain prime if $2p + 1$ is also prime. As an example, the number 5 is a Germain prime because the number 5 ($p = 5$) is prime having two distinct factors (1 and 5), and twice that number plus one ($2p + 1 = 2(5) + 1 = 11$) is also prime having two distinct factors (1 and 11). As a woman, Germain could not enroll in university and earn a degree, though her work was certainly worthy of such. In fact, Gauss, after learning she was a woman, petitioned to have an honorary degree awarded to her, but this was never done. She worked independently throughout her life as a mathematician, physicist, and philosopher, but due to sexism, her death certificate lists her only as a property owner. However, in 2003, the French Academy of Sciences established the Sophie Germain Prize in her honor.

Euphemia Lofton Haynes is the first Black woman to receive a PhD in mathematics. A 1943 graduate of Catholic University, she pushed beyond boundaries that had left many outside the discipline. She became a professor at the District of Columbia Teachers College and managed the Division of Mathematics and Business Education Department. Her story is empowering not only because of what she achieved academically but also because of how she used this success to propel others forward. She undertook a 47-year career of advocacy for students of color. Her work led to improvements in schools as she challenged the harmful effects of racial segregation. Haynes died in July 1980. She was 89.

These two women are remarkable in what they were able to achieve with the barriers set before them. They remind us too of the fact that there are individuals out there who have the potential to make outstanding contributions to mathematics and mathematics education if only they are given the opportunity, tools, and pathways to do so. Imagine how much we would lose out on as a society if these women did not break through the barriers! When women are marginalized, what advancements does society miss out on? Let's be clear: Women are still fighting for equality in mathematics and beyond. What are we missing out on as a society because so many are still fighting to get past the barriers instead of more easily being able to share their abilities with the world?

Resources: Women Mathematicians

Melfi, T. (Director). (2016). *Hidden figures* [Film]. Fox 2000 Pictures, Chernin Entertainment & Levantine Films.

Musielak, D. E. (2015). *Prime mystery: The life and mathematics of Sophie Germain*. AuthorHouse.

Neuenschwander, D. E. (2017). *Emmy Noether's wonderful theorem*. JHU Press.

Osen, L. M. (1975). *Women in mathematics*. MIT Press.

Riddle, L. (2022, July 5). *Biographies of women mathematicians*. Agnes Scott College. https://mathwomen.agnesscott.org/women/women.htm

Williams, T. (2018). *Power in numbers: The rebel women of mathematics*. Race Point Publishing.

World Women in Mathematics. (2018, November 19). *Journeys of women in mathematics* [Video]. YouTube. https://www.youtube.com/watch?v=uNJ7riiPHOY&ab_channel=WorldWomeninMathematics

COURSE PLACEMENT IN MATHEMATICS: THE CASE OF BLIND REFERRALS

It is not just women who are often marginalized in mathematics. Let us now look at one way Black and Latinx students are pushed out of the subject. Black and Latinx students are more likely to be placed into lower-level mathematics classes than their white peers in both elementary school and secondary school, meaning that they are frequently denied the opportunity to take advanced courses in high school that position them for acceptance into elite colleges and into STEM fields in general (Faulkner et al., 2019). Course placement even at early ages has a significant impact on students' opportunities as they age.

> We blunt our children's possibilities nearly from the start, telling far too many of them at a very early age that math isn't for them. Sometimes those communications are explicit; often they're embedded in decisions, by schools or districts, to put students on different tracks as early as third or fourth grade and teach them math that often limits how far they can go. Unbeknownst to the children or their families, these grouping decisions will decide the students' academic progress until the end of high school and beyond. This is far too early to make choices for students that can affect the arc of their lives. It is an unconscionable waste of human potential. (Boaler, 2022)

During school segregation in the 1960s, Black schools in North Carolina required their students to take and pass an algebra class to graduate from high school. At the time, schools serving white children did not. That is, Black students were being exposed to a more rigorous program of study in mathematics. However, when the schools desegregated, those same Black students were placed into lower-level mathematics classes than their white peers despite having previously taken more advanced coursework. The pattern continues to this day. National data from the United States indicates that with comparable academic records and socioeconomic standing the odds of a Black child being placed into Algebra 1 by eighth grade is two-thirds that of a white child (Faulkner et al., 2019). One of the reasons for this is that in many cases placement into advanced courses not only requires that students have a solid record of academic performance, but often also relies on teacher recommendations. Therefore, these students are subject to implicit bias.

I want to be clear that the individual teachers who are working with these students and making these referrals are not purposefully trying to disadvantage their Black students. I would imagine that as teachers they are interested in supporting and preparing all their students, as best they can. What I am referring to here is a set of implicit biases that we each carry with us due to our upbringing, our histories, our experiences, and the social world in which we exist. I have them, and so do you. We live in societies where being white affords one benefits that are not afforded to those with darker skin. It is to be expected that there will be biases around race that each of us carries but may not even be aware of.

> **It is to be expected that there will be biases around race that each of us carries but may not even be aware of.**

These biases drive our decisions, including those that affect students.

This implicit bias was made visible in the case of gifted program acceptance in one large, diverse school district. In this district, as in others, there exist gifted and talented programs at the elementary school level. These programs aim to identify talented youngsters who then are placed together as a cohort, taking their classes together and separate from the rest of the children in their grade at their school. These students get to participate in a more rigorous academic program than their peers. Further, they maintain their status as gifted students throughout their studies as there is no reassessment. To be identified as gifted, students take an examination. The examination is the same for all students who take it, but not all students in the district take the exam. Typically, teachers refer second graders for testing, and parents can also choose to have their child tested. David Card and Laura Giuliano (2016) examined data from the district to determine patterns and practices around identifying gifted students. With this referral-based screening system in place, 28% of the students who were placed into gifted programs were Black and Latinx (Card & Giuliano, 2016). Keep in mind that 60% of the students in the district identified as Black and Latinx. Thus, the percentage of Black and Latinx students in the program was far less than their representation in the district. The same school system later changed the way in which they identified students for the gifted program. They adopted universal screening. That is, the examination used to determine entry into gifted programs was now offered to all students in the district. Referrals were eliminated. The change in how students were identified led to a big shift in who was accepted into the program. Black student placement into the gifted program in this same district after the change to universal screening rose by 80%. The placement of Latinx students into the gifted program rose by 130%. Thus, Card and Giuliano contend that universal screening increases the number of Black and Latinx students in gifted programs. It is not that these students were not talented enough to get into the program before, but the screening process, which relied on referrals, allowed for implicit bias, and this narrowed the pool of qualified Black and Latinx students who took the exam and were admitted into the program. When universal screening was adopted, as expected in a society where being white affords you privileges not afforded to those whose skin color is darker, the percentage of whites in the program dropped, and the percentage of Black and Latinx students rose. This is without any change to the exam and its scoring. Further, the IQ scores for the newly identified students were very similar to those of the students who were accepted under the referral system. What changed wasn't the students or the criteria for acceptance, but rather the influence of implicit bias on the identification process. Five years after the universal screening was put into place, even with ample evidence that the newly identified students were just as capable and as successful as those who were accepted under the referral model, and even with clear data pointing to the fact that this was a more equitable system because it yielded a distribution of students by race that was more consistent with the makeup of the district in general, universal screening was abandoned as it was deemed too costly.

The mantra of blaming the continuation of racist policies in terms of student placement on budget constraints is one heard often. The short of it is that those tasked with creating budgets and allocating funds to schools have, over time, been neglectful of schools and programs that serve Black and Latinx students. It is another socially accepted belief that urban cities are strapped for funds and cuts are inevitable, but the practice of underfunding schools has been more readily applied in the recent past. More progressive systems of taxation and the elimination of tax breaks for large corporations would generate significant funding that could be used to sustain programs such as universal screening and to provide other supports needed to adequately serve Black and Latinx students. In Chapter 8, I deal more directly with the inequities in budgeting and the persistent underfunding of schools that serve Black and Latinx students.

The examples just described are a persistent reminder of the fact that historical realities and the continuing marginalization of all women as well as Black and Latinx individuals in mathematics continue to drive the reality that the discipline remains predominantly white and male. We all lose when mathematics is limited to some. We miss out on what could be valuable contributions by those who are pushed out of the field. People with different backgrounds bring new insights and differing perspectives to the discipline, thus enriching it and society. That students who are Black, Latinx, and women/girls (and all intersections thereof) struggle with mathematics is taken as a given, and as a result, decisions are made. Policies are instituted (or eliminated) based on our socially constructed views of who is *bad at math* even in the face of evidence to the contrary. To see a sizable increase in the number of women and people of color in mathematics consistent with their representation in our society will take a concerted effort at dismantling the systems of oppression prevalent in our society. Some of this is outside what an individual teacher or school leader can do, but there are steps that those in schools today can take in and out of the classroom.

Resources: Organizations Aimed at Broadening Access to Mathematics and Challenging Bias in K–12 Math Education

- The Algebra Project (algebra.org)

- Benjamin Banneker Association (bbamath.org)

- Lathisms: Latinx and Hispanics in the Mathematical Sciences (lathisms.org)

- Mathematically Gifted and Black (mathematicallygiftedandblack.com)

- TODOS: Mathematics for All (todos-math.org)

- The Young People's Project (typp.org)

WHAT CAN YOU DO TO CHALLENGE TRADITIONAL BELIEFS ABOUT MATHEMATICIANS?

Those who consistently portray individuals in mathematics as being brilliant must be pushed to provide alternate, more accurate depictions of mathematicians. When we see representations of mathematicians that are skewed toward socially awkward white men, let us remind ourselves of how narrow these are—how they fail to capture the full range of individuals who have and continue to make progress in this field—and let us consider how these characterizations keep diverse individuals with an interest in the subject from pursuing it because they do not see themselves reflected in society's characterizations of those who excel in the subject. A narrow view of who can excel in mathematics is limiting to every race, ethnicity, and gender as it keeps people who have the potential to make valuable mathematical contributions to society away from the field. As consumers of popular culture, let us reward those who create diverse mathematically able characters. Let us be consumers of media and culture that highlights women, Black and Latinx individuals, Native Americans, and those with disabilities, among others, as mathematically able so that these individuals begin to see themselves in this light as well.

> To challenge traditional views of who can do math in the classroom:
>
> - Read stories about diverse mathematicians to young children.
> - Have students research and create presentations on diverse mathematicians.
> - Watch films and TV episodes with your students that feature diverse mathematicians and scientists, and start a conversation with them about the programs.
> - Consider starting a math club for underrepresented groups.
> - Invite mathematicians or those in math-related fields to give talks to students at your school.
> - Decorate your classroom with representations of diverse mathematicians.
>
> For any of these ideas, there is ample information online. One place to start is "Links to Resources on Not Just White Dude Mathematicians" at https://arbitrarilyclose.com/links-to-resources-on-not-just-white-dude-mathematicians/

As educators, you can point out these narrow characterizations to students. Once these narrow characterizations become exposed, their ability to dominate the narrative is weakened. Instead, infuse your classrooms with representations that better reflect the diversity of the field and highlight the contributions of those too often obscured and erased. We saw in this chapter that textbooks represent the roles of girls and boys relative to mathematics in different ways. Let's point this out to students as well, for the students in our classes today are the textbook writers, novelists, and filmmakers of tomorrow. If they are aware of the skewed depictions in our texts and media today, they can not only be critical of them but also choose not to repeat them when they engage in careers in these and related fields.

We saw in this chapter, too, that programs aimed at marginalized groups in mathematics have the potential to change student views of the field. Let us work to expand accessibility to such programs. Consider starting a math club for girls or enrichment programs for those underrepresented in mathematics. Lastly, let us also support those marginalized in the field through positions of leadership where their voices and their actions can further support those like them.

Questions for Reflection

For Teachers

- What representations of mathematicians and mathematicians in training are present in your classroom? How can you work to make these more diverse?

- How might you start a conversation with students around depictions of mathematicians in the media and the textbooks that they use?

- In what ways can you bring in the stories that highlight the diversity of those who have made contributions to mathematics using films, texts, or other sources?

- What opportunities for students can you create to support women and other underrepresented groups in mathematics?

For Instructional Leaders

- What activities can you incorporate in your work with teachers to highlight the implicit biases we all have?

- What resources can you provide that challenge traditional views of mathematicians?

- In what ways can you support teachers as they work to open opportunities for those students from traditionally marginalized communities?

(Continued)

(Continued)

For Administrators

- What policies and methods are used in your school to identify students for enrichment opportunities and inclusion in advanced courses in mathematics? How can these be made to capture the diversity of students present in your schools?

- How can your hiring practices be modified to ensure that your faculty, administration, and staff represent the diversity present in our society?

- What ways can you use to ensure that the educational resources and curriculum adopted by your schools reflect the diversity of those who have contributed to mathematics?

- How can budgetary decisions be made to support the creation of programs and activities specifically geared to supporting underrepresented groups in mathematics?

CHAPTER 4

WE ARE ALL *MATH PEOPLE*

In this chapter we will:

- Redefine what is meant by the phrase *math people*.
- Consider the many ways that humans are seekers of patterns.
- Explore research on babies' recognitions of mathematical patterns.
- Consider the role of context in students' learning of mathematics.
- Reflect on how you can challenge traditional beliefs about math people.

DO *MATH PEOPLE* EXIST?

There are those who believe that some select individuals are born with a natural ability to do mathematics. These so-called *math people* somehow see their way through mathematical problems with ease, while the rest of the world—those without the magical math gene—struggle through it. The number of individuals who possess this gene is, from what we are led to believe, quite small. While some might argue against the existence of math people, I firmly believe they exist. What I reject is the idea that they are few in number and somehow possess mathematical abilities beyond the average person. They *are* the average person. As we have reconsidered what mathematics is and defined it as the study of patterns, it can be argued that we are all *math people*. Humans, after all, are inherently seekers of patterns.

> As we have reconsidered what mathematics is
> and defined it as the study of patterns, it can
> be argued that we are all *math people*. Humans,
> after all, are inherently seekers of patterns.

THE SEARCH FOR PATTERNS

One pattern that has been investigated at length is the relationship that exists between various distances and angles on a circle (arcs, diameters, circumferences, chords, inscribed angles, and so on). There are many relationships, though perhaps the most famous results in the mathematical constant pi, the ratio of a circle's circumference to its diameter, represented by the Greek letter π. The first hundred digits of pi are 3.1415926535 8979323846 2643383279 5028841971 6939937510 5820974944 5923078164 0628620899 8628034825 3421170679. The spaces here are for ease of reading only. The number was known to many ancient civilizations, and there is even a reference to it in the Bible in a discussion of a circular altar where it is approximated as 3. For centuries, mathematicians looked for patterns in the digits of pi. Some people devoted their entire lives to this endeavor. In the 1760s, Johann Heinrich Lambert proved that the number is irrational and as such there is no pattern to be found. Still, the search for patterns in numbers or elsewhere is a continuous human endeavor. My daughter was born on April 13, 2014. When written 4-13-14, the number is a palindrome. It reads the same backwards and forwards. Indeed, had she been born on any day from April 10 to April 19 that year, her birthday would have been a palindrome. As someone who values number patterns greatly, this little reality about her birthday brings a smile to my face.

We all search for patterns in our daily lives. Assume you have taken to going to the grocery store once a week. Consistently on Saturday afternoons, you make your way down the aisles and stock up on the ingredients needed to make your favorite dishes. This week, you end up leaving a bit later than usual and notice, when you get there, that the line to check out is much longer than it typically is. You start to think through why that might be, coming up with some plausible explanations for it, and you remind yourself to leave home a bit earlier the following week. This long line breaks with a pattern you have noticed. Perhaps you don't much like cooking and instead eat out regularly at your favorite restaurant. You start to notice that your favorite waitperson seems to be there most often on Mondays, Tuesdays, and Wednesdays and is almost always assigned to the tables in the back left of the establishment; so, when you remember, you request a table in the back and show up early in the week. After all, the probability that you end up being served by your favorite waitperson is much greater that way.

While these examples are quite trivial, there are a myriad of situations where being a seeker of patterns is quite useful. This is the case for scientists combing through data to determine the best way to stop the spread of cancer or nutritionists reading through labels to make the best food recommendations for their patients. It is the case for financial advisors who try to find investment options that may yield vastly lucrative results for their clients and thus for themselves as well. Patterns help us to design better cars, develop more graceful dance routines, determine where and how to market products, and decide which candidates to support in an upcoming election.

Children's Recognition of Mathematical Patterns

Humans, again, are natural-born pattern finders. Even the young among us can see the patterns around them. When my daughter was six, she encouraged me to exercise, so I printed out a calendar and hung it on the door of my closet. My daughter excitedly put a sticker in the box for each day I exercised. This incentive chart was quite useful, but more than anything I was motivated by the smile on her face each time she placed a sticker. She would comb through the stickers she had, trying to pick out the right one, and do her best to put it on as straight as possible. On the day I am recalling, it was a five-point star. She put it on the page, declaring she had "put it on right" as she stepped away. Then, after a pause, she said that it would still be right even if she turned it a bit. Not one to pass up an opportunity for a math lesson, I asked her what she meant. She explained that she put the star with "this part up" but that if she turned it and put it with this other part up it would still look the same. The parts she was referring to were the points of the star. She was then able to explain that since the star she stuck to the page had a point at the top, turning it so that any of the five points were at the top would make the star look as it did when she affixed it to the page. Without the mathematical language, what she was talking about was rotational symmetry. The star has five points, and thus any turn of 360 ÷ 5 = 72 degrees will yield a star that looks the same as the original, as will turns where the number of degrees is a multiple of 72.

At my daughter's preschool, teachers routinely worked with children around the concept of sorting. Teachers gave students round objects—for example, balls of all different kinds, including cotton balls. They asked students to describe the properties of the balls. Some of the students talked about size, noting that some were big while others were small. Other students highlighted the color of the balls. Students talked about texture, noting those that were soft, smooth, rough, or spiky. Additionally, some talked about the use of the objects, noting some of the balls were for sports while others were not. In this example, toddlers were able to create sets of objects that share certain properties. For example, they picked out all the balls that were blue. While it might not seem like mathematics to some, the ability to identify properties of objects and sort by such properties is a mathematical skill. Going forward, one might sort shapes, solids, knots, functions, or any number of mathematical objects based on their properties.

A child sees mathematical patterns the way someone with more mathematical training would; the child just may not yet have the formal language to explain them. On another occasion, my daughter, then just a toddler, was playing with Magna-Tiles® on the floor. These are a set of plastic building pieces produced by Valtech that have magnets along the edges that allow one to build complex three-dimensional structures. On this day, my daughter was laying shapes flat on the floor. She took some squares and placed them next to one another. The pieces snapped together due to the magnets. She did this with squares and then triangles. In each case, she covered an area on the floor without leaving any gaps. I noticed this and wanted to explore with her which shapes would be able to cover the floor without leaving gaps (tesselate the plane) and which would not be able to. I asked my daughter if she noticed that when she used the squares there were no spaces where we could see the floor, but she was too excited by what she was doing to entertain my questions. I kept at it for a bit, eventually recognizing that my persistence was not going to lead to a discussion, and let her play undisturbed. When she had finished every single shape she had in her box, she looked at her creation, very impressed with her work. I asked her what she saw. She responded, "Antarctica." In time she would explain that the areas where the pieces completely covered the floor were ice and that the places where the floor could be seen represented the water that is normally hidden under the ice but is visible when there are gaps. She did notice which shapes made ice and which left patches of water (as I learned from further discussion), but where I saw tessellations, she saw Antarctica. Once she was excited about sharing the work she had done, and I understood she saw the shapes as pieces within patches of ice—we had a common language—we were able to talk a bit about how some shapes cover the floor and others do not.

Mathematics as Part of Babies' Nature

In 1992, Karen Wynn demonstrated the fact that mathematics is part of our nature by studying how babies of only four months in age respond to situations involving the quantities of one, two, and three. Wynn showed babies a stage on which there was one puppet. Then a screen would cover the stage so the puppet could no longer be seen. Someone would walk over to the stage and place a puppet over the screen onto the stage. At this point the babies would expect to see two puppets when the screen was removed. However, the researchers would sometimes alter the number of puppets behind the screen (adding or subtracting) without the babies being able to see this. If there were two puppets (as expected), the babies would look at them for a time and then divert their eyes as usual, but if the number of puppets didn't match the expectation, the babies would stare for a much longer period of time. This happened over and over again. The research showed that babies have an intuitive sense of quantity at least as far as one, two, and three are concerned. They certainly are too young to know what the number *one* is or to know that this number represents the quality of oneness, but still in some intuitive way the babies in Wynn's study knew that one puppet and one puppet should yield two puppets. They were also able to accurately expect what would occur if puppets were removed, so there was an intuitive sense of quantity, as well as the addition and the subtraction of such quantities, again for one, two, and three. The babies were unable to make these determinations if the number of puppets was larger than three, no longer consistently fixing their gaze for a longer period of time when there was an unexpected number of dolls. Subsequent research refined the study to ensure that babies weren't reacting to some other facet of the situation— for example, the placement of the puppets rather than the number of the puppets. These experiments and variations on them have been replicated all over the world, and the results remain the same. Further research has shown that the results remain consistent even when the children are but a few days old and that in addition to seeing differences in one, two, and three objects, babies can recognize differences in the number of times they hear a particular sound. It seems that as a species, we are hardwired for mathematics.

Patterns in Our Daily Lives

If we're so hardwired for mathematics, how is it that the natural pattern seekers of yesterday grow to be the mathematically fearful students and adults of today?

> If we're so hardwired for mathematics, how is it that the natural pattern seekers of yesterday grow to be the mathematically fearful students and adults of today?

How is it that even those who enjoy patterns, logic, games, and puzzles will claim to be *bad at math*? One of the reasons has to do with how most of us learned mathematics in school, which we will discuss further in Chapter 6. Another is that the natural pattern seekers of yesterday do grow up and continue to be pattern seekers into old age. However, our definition of what counts as mathematics, reinforced through years of formal schooling, does not make room for much of what pattern seekers do in their daily lives. As a result, we neglect to see much of the mathematics that we are, in fact, engaged in. When we head out to do errands and plan the best route (whatever our definition of *best*), we are engaged in mathematics. We engage in mathematics when we estimate the amount of money we are spending at the store or take the time to discover the dollar amount for a 20% discount or a 15% tip. When we pack items into our truck, rotating bags and boxes or moving things from one side to the other to ensure that all fits, we are again engaged in a mathematical task. The same is true when we play a game of Tetris or Sudoku, or when some folks take items from the shelf at the supermarket and arrange them in the cart so everything has its place. These last few are examples of packing problems and have been studied by mathematicians extensively.

When we change the angle of the bat and adjust the speed at which we are swinging to meet the ball, we are engaged in mathematics, though not with paper and a pencil. The same is true if we think, instead of an outfielder running in an arc to meet a fly ball, of a soccer player who kicks the ball at a certain angle to produce a spin that curves it into the corner of the net, a basketball player shooting the ball in an arc toward the hoop, and a tennis player moving their racquet in such a way as to hit the ball with top spin so that despite the force with which they hit it, the ball lands in the court. Those of us who play sports, from football to soccer to tennis to golf and more, are all making mathematical calculations and adjustments in our minds and bodies with the aim of improving our chances of winning. We are also playing on fields, courses, and courts covered in shapes and using our knowledge of these throughout the game. Those who enjoy watching sports are often immersed in numerous statistics about the game they are watching and the teams they follow. Those who build and guide teams to victory in fantasy leagues are using probability to make the most use of their budgets in acquiring players and get the most out of their players on game days throughout the season.

Designing a garden involves shapes, considerations about how much soil one needs, and decisions about how far to space out various plants. Painting involves determining the amount of paint needed and knowledge of shapes depending on what one is painting. Knitting, sewing, and quilting all rely on mathematics as do many of the patterns that textile workers use. In fact, there are some beautifully intricate mathematical designs full of reflections, rotations, and symmetry in some of their work. Those who play or write music, as well as those who sing or dance, rely on the mathematical relationships between notes and chords. Music is mathematics. Similarly, hobbies such as puzzles of all kinds also rely on mathematics whether to fit pieces into spaces, rotating them as needed; to work with numeric patterns; or to use logical reasoning to determine what five-letter word fits into a specific space that has an *m* as the third letter and fits the definition given. Despite the many ways that we engage in mathematical processes, many people often do not consider themselves mathematical creatures. Partly this is because our definition of what counts as mathematics is a narrow one.

> Despite the many ways that we engage in mathematical processes, many people often do not consider themselves mathematical creatures. Partly this is because our definition of what counts as mathematics is a narrow one.

It is also because of the ways in which many view themselves with respect to the discipline. We have been taught in ways that perhaps did not strengthen our understanding of the subject and that did not develop in most a love for it, within a culture that allows for one to admit one is *bad at math*. It, therefore, becomes easy for many to accept the belief that they are not good at it, that they will never be good at it, and that, as such, they are not mathematical in the ways in which they interact with the world. Yet each of us is mathematical in the ways we interact with the world daily. If we define mathematics broadly, it is difficult to think of activities that one engages in that do not have a mathematical component. I urge you to try and identify something you did today that did not involve mathematics in any way, even if subconsciously.

To highlight that mathematics is all around us in (and out of) the classroom:

- Send students on a hunt to find the mathematics all around them.

- Have students tell, write, or draw about the mathematical experiences in their everyday lives.

- Encourage young children to search for shapes while on a walk around the community or the school.

- Invite older students to bring in examples of math in the news and/or in their community.

NATURE IS THE BEST MATHEMATICIAN

In his book *The Math Instinct: Why You're a Mathematical Genius (Along With Lobsters, Birds, Cats, and Dogs)*, Keith Devlin (2005) states that nature is the best mathematician and then recounts with detail the ways in which various animals—ants, bees, dogs, owls, and more—undertake ways of being that can be described using mathematics. He gives the example of the mathematician Tim Pennings, who was walking his dog, Elvis, along the beach and tossing a ball into the water for the dog to catch. He noticed that if he threw the ball straight into the water, Elvis would swim right toward it in a straight line. However, if he tossed it diagonally into the water, the dog would not immediately jump into the water and swim toward it. Instead, Elvis would run alongside the water's edge for a bit and then turn toward the water and swim for it in a straight line from there. The dog can run along the water's edge faster than he can swim into it; thus, this method gets him to the ball faster than if he immediately threw himself into the water, but there is a decision to be made about how far to run before heading in. The problem can be modeled using calculus, and if one knows the speed at which the dog runs and swims as well as the various distances involved, it is possible to determine the best choice for when to turn into the water. By *best choice*, I mean the point at which we minimize the time it takes to get to the ball. After observing this behavior in his dog, Pennings decided to head out to the beach with his tape measure and his stopwatch. He threw the ball diagonally into the water and chased his dog, noting the point at which he turned into the water. His dog runs faster than he does, but his swimming is solid and so he was able to catch up to the dog by the time he got to the ball. Pennings was therefore able to compute the distance that the dog had traveled in the water and on land as well as the time it took the dog to do so. He undertook the process over and over again, taking measurements until he tired himself out. With all the measurements in hand, he got to work on analyzing the dog's choice of where to turn into the water and found, remarkably, that the dog tended to choose the ideal spot to turn or close to it. Now certainly the dog was not doing calculus to determine where to turn, but rather instinctively turning at the right point for the most efficiency. The dog's nature is such that the mathematics is hardwired. It is this way with humans as well.

WHY CONTEXT MATTERS

The context in which we engage in mathematics plays a big role not just in whether we see the mathematics we are doing as mathematics, but also in whether we can carry out the calculations at all. When instinctively running to the ball, Elvis was able to determine the best spot to turn into the water, but if given a diagram of the problem, a pencil, and some paper, he would not have been successful in making the determination. He is, after all, a dog. It turns out that people are this way too. While we may be engaged in all sorts of mathematical behaviors and in some ways instinctively managing to figure out the math required, if those same calculations and questions are placed on paper in front of us, we may falter at being able to answer them. That is, you might be able to determine the best route in practice, but stripped of the context and your list of errands, this may prove a challenging problem indeed.

This was made clear with the work of Terezinha Nunes Carraher and her colleagues (1985) who studied street vendors in Brazil. They noticed that the vendors were able to do computations in their work involving the cost of multiple items, computing totals, and giving change even when there were multiple items involved and differing prices. The way they worked out the math differed from the traditional paper-and-pencil algorithms taught in schools, which the researchers learned by interviewing the vendors about how they were determining the various amounts. They had the numerical fluency needed for the type of job they were performing and relied on nonstandard algorithms. As an example, if someone was buying three items costing 16 *reales* each, instead of the traditional algorithm where one lines up the 3 under the 6, multiplies these, and carries, the vendor might use one of the nonstandard approaches listed in Figure 4.1.

Figure 4.1 • *Standard and Nonstandard Approaches to Multiplication*

Standard Approach	Nonstandard Approaches
16 × 3 —— 48	Separate the 16 into 10 and 6, multiply each of those by 3 (30 and 18), and then add to obtain 30 + 18 = 48.
	Find 15 times 3 to obtain 45, which may be a more comfortable problem to start with. Then realize you are short one quantity of three and add 3 to the 45 to obtain 48.
	Separate 16 into 10 and 6. Count three sixes (6, 12, 18) followed by three tens (28, 38, 48) to arrive at 48.

What is interesting is that while the street vendors were incredibly adept at these calculations when working, they were much less successful at carrying out the same calculations when they were asked to do so on paper-and-pencil assessments where they tended to keep to the standard algorithm. This was true whether the questions were stripped of context or whether they were presented in the context of buying and selling goods just as they were doing on the street in their work.

Resource: Procedural Fluency

Bay-Williams, J. M., & SanGiovanni, J. J. (2021). *Figuring out fluency in mathematics teaching and learning, Grades K–8: Moving beyond basic facts and memorization*. Corwin.

We, too, engage in mathematics daily that we may fail to recognize as mathematics. Most of us were taught only the traditional algorithms for the four basic operations, yet with time and exposure we gained the ability to think about adding, subtracting, multiplying, and dividing numbers in ways similar to the street vendors and to use this in our daily lives. Presently, young children are introduced to various algorithms to build this fluency right from the start. This fluency is more easily attained when the mathematics we do is connected to our lived experiences. We need to find ways to connect the mathematics that we do daily with the mathematics that we do in the classroom as well. To be more aware of the connections between what we are learning and our lives makes the content more real and easier to understand. Context matters. When we teach mathematics in ways that do not connect to students' lives, it makes it hard for them both to be excited about the content and to make sense of it. Is it not much easier to read a book on something you enjoy than one not connected to your interests or experiences at all? You can develop your ability to read and write by reading a variety of texts, but you might get more out of it if the readings connect to you in some way. With mathematics it is the same. You might get more out of the lesson if it connects with your experiences and history in some way. Connecting with students' experiences and cultures is part of culturally relevant teaching, which we will explore further in Chapter 5.

We engage in mathematics regularly, whether considering the angle at which we enter a parking spot or the direction in which we move our glove to catch a ball. While we certainly aren't standing on the curb with a protractor or on the field with a tape measure, we are using mathematical realities to accomplish these tasks. That is, we are *math people*—all of us. One way to make this clear is to find ways to highlight the mathematics in our everyday lives and share this with our students and others so they become more aware of it and find it harder to dismiss their natural mathematical abilities.

HOW CAN YOU DISMANTLE HARMFUL BELIEFS ABOUT MATH PEOPLE?

- Develop in students the belief that they are math people, as are you.

- Connect mathematics to contexts that hold meaning for students.

- Create opportunities for students to identify mathematics around them.

Questions for Reflection

For Teachers

- How can you instill in your students the reality that they are math people?

- In what ways can you engage students in the exploration of the patterns around them?

- How can you connect the content you teach to the contexts that matter in students' lives?

For Instructional Leaders

- In what ways do you provide opportunities for teachers to explore patterns as part of their professional growth?

- How can you work with teachers around leveraging real-world contexts in the classroom?

- What can you do to support the idea that all of us, students and teachers alike, are math people in your work with teachers?

For Administrators

- How do the resources and curriculum you adopt support the idea that we are all math people?

- What opportunities do you provide that allow those in your school to share the many ways that they use mathematics in their daily lives and build on these in their work?

- How do or might you leverage the expertise of those in your community (both within and outside the school) as doers of mathematics?

CHAPTER 5

IDENTITY IN MATHEMATICS EDUCATION

In this chapter we will:

- Consider the role of identity in mathematics education.

- Explore what is meant by mathematics identity.

- Examine research on the connections between identity and achievement in mathematics.

- Consider the role of story problems in mathematics.

- Reflect on how you can build on students' identities in your teaching.

IDENTITY INSIDE THE CLASSROOM

The fact that identity impacts student learning inside the classroom is well established. Students are more likely to be engaged when the material they are learning values their histories and experiences. When students see themselves reflected in the texts they read and the content they study, they are more likely to connect with that material and thus better learn it. Yet, with respect to mathematics, it is often believed that identity does not play as strong a role. After all, it is widely believed that mathematics is neutral and that the teaching of mathematics is an exercise in objectivism. Both of those are far from the truth.

Many of the mathematics textbooks and materials we use are written with a white, middle-class student in mind. As such, they do not often reflect the realities and experiences of Black, Latinx, and Indigenous students; those who are recent immigrants; or those from less affluent backgrounds.

This makes it hard for students to relate to the content and make sense of the mathematics. William F. Tate IV has studied cultural contexts and assumptions made in mathematics classrooms. Tate (2005) presents an example where students were asked to determine which was a better price: buying a card to ride the metro, in which you pay for each ride, or buying one that allows you to ride the metro as often as you'd like for the entire week. The cost per ride and the cost of a weekly card were given. Students were told that their parent would be using the card to commute to and from work. The *correct* answer was that the "pay as you go" option was least expensive and thus the better price. While many of the students in the class were getting the problem wrong, upon closer examination Tate was able to determine that some of these students were doing the computation correctly but choosing the weekly card for reasons that had to do with their own experiences. The problem writers assumed that the parent in question would be traveling to and from work 5 days a week for a total of 10 trips per week. Students in the class, however, had parents who worked multiple jobs or who worked more than 5 days a week and thus traveled to and from work more than 10 times in a week. Further, in many cases the cards were shared among members of the family. Therefore, a parent might pass the card to a child who would use it as well, adding to the total number of trips that the card would be used for. The issue here was not with the mathematics but with the context that was used, which centered not on the lived experiences of the students in the class but rather on a white, middle-class experience.

In a similar example, Tate (2005) noticed that students in a math class were not engaged with the lessons that centered on dividing up a pumpkin pie. It was around Thanksgiving, and using pumpkin pie was a way for the teacher to try and connect the content to the lives of students. The issue is that some of the students had never heard of pumpkin pie and so the problem seemed foreign to them. Tate spoke to these students about what kind of desserts they enjoyed with their families and found that by changing the dessert to sweet potato pie some of the students were able to connect with the context and were more willing to engage with the mathematics as a result. In short, the context matters. It should be noted that no curriculum can accurately and effectively reflect all cultures and identities. Often, attempts to do so become superficial and over-rely on stereotypes. Teachers must also know their students well and be supported in adapting curricular materials to their students' contexts. But it should not be either/or. The ideal combination includes resources that have better representation of race, gender, and ability being used fluently by teachers who know their students well and are trained and equipped to use those materials to support their students' full identities. The bottom line is that valuing the experiences and histories of students with marginalized identities helps those students better engage with the material. The problem of context goes beyond individual teachers in individual classrooms.

> ### Resources: Culturally Responsive Teaching
>
> Gay, G. (2018). *Culturally responsive teaching: Theory, research, and practice.* Teachers College Press.
>
> Hammond, Z. (2014). *Culturally responsive teaching and the brain: Promoting authentic engagement and rigor among culturally and linguistically diverse students.* Corwin.
>
> Matthews, L. E., Jones, S. M., & Parker, Y. A. (2022). *Engaging in culturally relevant mathematics tasks: Fostering hope in the elementary classroom.* Corwin.
>
> Matthews, L. E., Jones, S. M., & Parker, Y. A. (2023). *Engaging in culturally relevant mathematics tasks: Fostering hope in the secondary classroom.* Corwin.

MATHEMATICS IDENTITY

Danny Martin defines mathematics identity as consisting of "the dispositions and deeply held beliefs that individuals develop about their ability to participate and perform effectively in mathematical contexts and to use mathematics to change the conditions of their lives" and "a person's self-understanding and how others see [the person] in the context of doing mathematics" (Martin, 2009, pp. 136–137). How we see or do not see ourselves as doers of mathematics affects our ability to engage in mathematical work and to achieve success in the discipline.

> **How we see or do not see ourselves as doers of mathematics affects our ability to engage in mathematical work and to achieve success in the discipline.**

Students who see themselves as capable mathematically are more likely to excel, in part, because that belief will result in their engaging in behaviors that are linked with success such as studying and completing their homework. Individuals who do not believe they can be successful mathematically are more likely not to engage in these behaviors, consistently and incorrectly assuming they are not likely to help. This reality is known as the *self-fulfilling prophecy*, a term coined by Robert K. Merton in 1948. Mathematics identity is also impacted by stereotype threat, especially among women and other traditionally underrepresented groups in mathematics. Stereotype threat is the fear that one will act in ways that reinforce negative stereotypes about one's group (Steele, 2006). For example, there is a negative stereotype of Black and Latinx people as nonintellectual, which students from these groups may fear reproducing. This may lead to students performing poorly because the fear affects the students' ability to engage in the work. It is challenging to perform at one's best in the presence of fear.

MATH IDENTITY AND ACHIEVEMENT

In a research project my colleagues and I engaged in, we investigated the relationship between mathematics identity and achievement for Black and Latinx secondary school students using data from the National Center for Education Statistics' (NCES) High School Longitudinal Study (HSLS) of 2009. The NCES is a federal entity for collecting and analyzing data related to education in the United States. The baseline survey was administered to 23,000 ninth-grade students in 2009. The sample included 944 schools and sampled public, private, and Catholic schools with ninth-grade students randomly selected from within each school. Follow-ups began in 2012 when most of the students were in 11th grade. High school transcripts were collected in 2013, and a last follow-up occurred in 2016, three years past graduation, when many of the participants were either in college or employed, having never attended college.

We focused on students who identified as Black and Latinx, employing hierarchical regression modeling to examine the relative impact of math identity, demographic variables, and school/parent social capital variables on the math grade point averages (GPAs) of the students in 11th grade. Hierarchical regression modeling allows researchers to determine if certain variables explain a statistically significant amount of variance in a given variable (for us, math achievement) after accounting for all other variables. The HSLS includes numerous questions that speak to a student's mathematics identity including questions that ask whether students see themselves as a math person, whether others see them as a math person, how often students think they understand their math assignments, and how confident they are about their ability to do well in their assignments and in their math class in general. We put these together to create a composite variable measuring math identity. The fact that identity and mathematics achievement are linked came through in the data, and it did so in a way that is extraordinarily hopeful.

Specifically, what we found is that mathematics identity—more than any of the other variables we examined—predicted success in secondary school mathematics as measured by the students' math GPAs in 11th grade. The variable *mathematics identity* was positively associated with math GPA across all the models that we ran. Further, this held true even when we controlled for other variables such as socioeconomic standing (a standardized variable established as a continuous index score based on a combination of income, education, and occupational prestige) and parents' social capital (a composite variable meant to measure the social resources that individuals can rely on to further their position, which may include relationships with individuals and organizations as well as knowledge of social norms, educational practices, and political systems). The results of this work were published in two papers that contain a more complete discussion of the methods employed and the results obtained (see Gonzalez et al., 2020, 2022). One additional result of the research discussed is that Latinx students performed better mathematically in schools where a higher percentage of students took Advanced Placement (AP) classes. Interestingly, this held true for Latinx students in general, as well as for Latinx boys, but not for Latinx girls. In all these cases, it is the percentage of students taking such classes and not the number of courses offered that is statistically significant. That is, the more students there are making use of these enriched academic courses, the better. Schools with more access to academic enrichment tend to serve Black and Latinx students better than those with a focus on remediation.

> **Schools with more access to academic enrichment tend to serve Black and Latinx students better than those with a focus on remediation.**

To increase focus on mathematics identity in the classroom:

- Have students write a math biography or do a recording where they describe their experiences with math.

- Consider using journal prompts in class to learn more about your students.

- Interview families to learn more about them (this can be done with surveys or formal interviews).

- Learn about traditions your students keep or holidays they celebrate, and incorporate those into your classroom.

- Use students' histories and experiences as the contexts for problems you solve in class.

IDENTITY AND STORY PROBLEMS

Students who are from groups historically marginalized in mathematics may struggle to build a positive mathematics identity as a result of social and historical realities and may suffer in what Gutiérrez (2017) calls *dehumanizing* mathematics education—that is, education that does not value students' histories, cultures, and experiences. It is education that expects students to leave who they are (their culture, experiences, family celebrations, and the like) at the door. The misconception that the teaching of mathematics is objective leads to the additional misconception that issues of identity need not be important considerations when teaching mathematics, but we have seen that this is not the case. Identity has a profound effect on the teaching and learning of mathematics.

Further, while the idea that student identity need not be considered in teaching mathematics is a misconception, it is not the case that we have been neglecting identity. To say that the books and materials used in the teaching and learning of mathematics are objective is completely erroneous. These materials are centered on the identities and experiences of white, middle-class, cisgendered students, the so-called mainstream students. Standardized exams that are used as gatekeepers to further opportunities are built around the experiences of these students as well. The identities of white, middle-class students are being attended to. The idea that identity is not a necessary consideration in the teaching of mathematics serves to obscure the reality that identity is in fact being attended to, though not for all students.

> **The idea that identity is not a necessary consideration in the teaching of mathematics serves to obscure the reality that identity is in fact being attended to, though not for all students.**

The identities of our Black and Latinx students are not being considered, celebrated, and built upon, unlike those of their white counterparts. Anita Bright (2016), in her article "The Problem With Story Problems," discusses some of the issues with the contexts used in mathematics problems from elementary school to calculus. She indicates that some of the most common themes involve consumerism where the frame of reference is the middle or upper middle class. She presents as an example the following problem from the fifth edition of *Precalculus: Mathematics for Calculus* (Stewart et al., 2005):

> Craig is saving to buy a vacation home. He inherits some money from a wealthy uncle, then combines this with the $22,000 he has already saved and doubles the total in a lucky investment. He ends up with $134,000, just enough to buy a cabin on the lake. How much did he inherit?

This problem is centered on the experiences of someone who can afford a vacation home and further who has rich family from which to inherit money. Other problems discuss inheriting gems, having multiple horses in one's corral, and arranging parking for a yacht (Stewart et al., 2005). In yet another example from a text for Grade 2 (Snider et al., 1999), there is a problem involving a "Hawaiian Dream Vacation" that includes the cost of various luxuries such as chartering a plane. Still another problem Bright (2016) presents involves renting a jet ski, and for the students who were using the text (Boswell & Larson, 2010), the cost was more than some of their parents were making in a week working a full-time job.

That many students cannot afford such luxuries, yet they are presented as if commonplace in these types of questions, is troubling as it marginalizes those who cannot relate to these contexts. Instead, questions and contexts could be crafted around the experiences of the students in your classroom so that the work connects with them in some way. This might make the material more relevant to the students' lives and promote greater engagement with the mathematics. Admittedly, this is a difficult process and requires a lot of work on the teachers' part. To create problems centered on our students, we must come to know them well and spend time recontextualizing mathematics so it is centered on them. This is not easy work. Yet we cannot take for granted that who we are—our fundamental identity—affects our learning, including the learning of mathematics. We need to attend to issues of identity within the mathematics classroom, as well as in the educational system in general, if we are to meet the needs of our students and truly support them as they grow in their understanding of mathematics.

ATTENDING TO IDENTITY IN THE CLASSROOM

In this chapter we have seen the role that identity plays in mathematics learning and its connection to achievement, especially for Black and Latinx students. To support students' achievement, we must attend to issues of identity in the classroom by centering the curriculum on their experiences. This means developing curriculum and supporting materials that honor the cultures and histories of the students in our classrooms and using the relevant materials that do exist. As a simple exercise, young students can study the flags of the countries their families or ancestors came from and identify the various shapes and colors represented. They can consider what colors appear most or least often. Older students can take those same flags and determine the area of certain shapes, or the area covered by certain colors. For much younger children, teachers can bring in picture books with mathematical themes that value diversity—perhaps because they are inclusive in the characters they portray or in the authors and illustrators who have written and illustrated them, or because they share stories of diverse cultures. Mathematics can be taught using contexts that speak to the students in our diverse classes.

Questions for Reflection

For Teachers

- How can you learn more about your students and their families?

- In what ways can you attend to identity in your classroom?

- How can you use students' histories and experiences as contexts in mathematical problems?

- In what ways can you incorporate culturally relevant teaching in your practice?

For Instructional Leaders

- In what ways can you support teachers in attending to student identity in their teaching?

- How can you attend to the identities of the teachers you work with?

- In what ways can you encourage culturally relevant teaching among the teachers you work with?

For Administrators

- How does your school value and celebrate the cultures and histories of its students, faculty, and staff?

- In what ways can you support teaching that is centered on students' experiences?

- How does the curriculum you adopt reflect the diversity of cultures represented in your school?

- How do you engage with families and community members?

CHAPTER 6

SCHOOL MATHEMATICS

In this chapter we will:

- Consider how mathematics traditionally has been taught.

- Explore pedagogical practices that challenge traditional views of what mathematics is and who can excel at it.

- Reconsider school mathematics to better support student learning.

- Reflect on how you can challenge traditional beliefs about mathematics while engaging students in the meaningful study of mathematics.

WHERE DO TRADITIONAL BELIEFS ABOUT MATHEMATICS COME FROM?

The socially constructed beliefs we as a society have about mathematics and mathematics education come from somewhere. We learn and have been learning these throughout our lives. One does not instinctively decide that it is acceptable to say one is not good at math. Instead, one hears other people say it over and over again and in time believes it is okay to say it as well. Similarly, the beliefs people hold about what mathematics is have been learned. While these lessons are reinforced throughout our lives by those around us, in the media, and elsewhere, many of the ideas people have about mathematics come from their experiences with school mathematics. We cannot improve mathematics instruction and challenge our harmful beliefs about mathematics and mathematics education by focusing on the classroom alone, however, for these are intertwined with the social realities of our world. Yet, we cannot address them without considering what mathematics was like for most of us inside classrooms, what it continues to be like for many, and the ways in which we can reconceptualize school mathematics experiences so that the subject comes alive for students in ways that for many it has not done in the past.

Conceptual vs. Procedural Understanding

One of the biggest problems with school mathematics is its reliance on the teaching of procedures. Often, students are presented with a set of problems and then a method for solving them that involves carrying out an algorithm—a set of steps. In fact, in a study of students taught using traditional approaches, it was found that when students were asked what mathematics is, the majority replied that it is about rules and procedures (Boaler, 2002). In their work, Stigler and Hiebert (2019) consider the teaching practices of teachers across numerous countries. They have found that teaching is a cultural activity that varies from country to country depending on the beliefs held in that country around the acts of teaching and learning. For too long, teaching across North America has relied on a script where the teacher presents a topic, does some sample problems, and then asks students to replicate the work done by completing very similar problems, usually on their own. Procedures are often taught without considering why they work or in what cases, if any, they might fail to work. Further, students are often passively listening and watching rather than engaging with the content in an active way. With a focus on procedural understanding rather than conceptual understanding, students may be able to solve problems, but they don't really understand the mathematics behind their work.

> **With a focus on procedural understanding rather than conceptual understanding, students may be able to solve problems, but they don't really understand the mathematics behind their work.**

Accumulating many algorithms in their mind without a true understanding of how they work, where they come from, or how they are connected is a recipe for disaster. In time, one forgets the steps and where to apply them or confuses one algorithm with another. It is hard to keep straight a set of seemingly distinct algorithms. Does this ring true for the classes you took growing up? How often do you see math courses conducted in this way today, even if not your own?

For various summers I taught a workshop for students who had failed the mathematics portion of the City University of New York (CUNY) entrance exam. Without passing this exam, the students could not enroll in courses at the colleges in the CUNY system. I found extremely eager students who, in most cases, could partially recall rules and procedures they had learned in their mathematics classes or who could recall the correct procedure but applied it to the wrong type of problem. "Two negatives make a positive," they would tell me, not knowing in which of the four operations this held true. Or they would say that when there are fractions involved one must cross multiply, not recognizing that this doesn't hold across all cases. Much of my work with them was a matter of clarifying the vast amount of information they had tried to keep straight in their minds, especially after not having used it for some time. When we broke down the procedures, discussing why they worked and how they connected to one another, their eyes lit up, and it made sense to them. Sadly, this doesn't always happen in K–12 mathematics classes. Often it feels like formulas drop from the sky. Students don't know where they come from or why they work but are expected to memorize and use them in very prescribed ways. When they stop using them, they forget them. Mathematics doesn't make sense this way because we never get to understanding the *why* of it all, remaining perpetually stuck on the *how*.

Passive vs. Active Learning

Another issue with the way that mathematics is traditionally taught is the reliance on passive learning—that is, relying on students to learn by watching and not, necessarily, by actively grappling with ideas. My father was a very handy man. He laid down the floors in almost every room of the house I grew up in, installed the kitchen cabinets, and redid the bathrooms, among other home improvement projects. As a child, I often helped him, but mostly this meant that I held the flashlight while he did the real work. As a result, I had an idea of how to do many projects around the house but no real experience doing them. This became clear when I moved into my own apartment and attempted to install a smoke alarm. I knew enough to determine what the wall was made of and find the tools needed for the job. Yet, not having ever held the drill in my hand before, I did not know how to apply pressure correctly, and the screw I was trying to drill into the wall flew clear across the room. Watching someone do something, however many times you do it, is not the same as doing it yourself. Students of mathematics, especially at the secondary level, know this all too well when they find the material makes sense in class but are baffled by the homework. There needs to be less of an emphasis on students watching teachers model procedures and more of an emphasis on students engaging in mathematics building themselves if we are to support students in their learning of mathematics. If students are to really know how to do mathematics, they must *build* mathematics and do so often.

> If students are to really know how to do mathematics, they must *build* mathematics and do so often.

MOVING BEYOND TRADITIONAL TEACHING: THE PRACTICE STANDARDS

In 2009, the United States introduced the Common Core State Standards Initiative as an attempt to have some level of national uniformity about what students should learn by the end of each grade level. Though there was a lot of political discussion around their adoption, the vast majority of states have adopted and continue to use the Common Core State Standards for Mathematics (CCSS-M). As shown in Figure 6.1, these include a set of eight practice standards. No longer is it enough to specify the mathematics to be learned; now there exists explicit language focused on *how* mathematics will be done to move away from rote learning, procedural understanding, and passive classrooms, and to instead move toward active learning, exploration, communication, and an emphasis on conceptual understanding.

Figure 6.1 • *The Common Core State Standards Initiative Mathematical Practice Standards*

- Make sense of problems and persevere in solving them.
- Reason abstractly and quantitatively.
- Construct viable arguments and critique the reasoning of others.
- Model with mathematics.
- Use appropriate tools strategically.
- Attend to precision.
- Look for and make use of structure.
- Look for and express regularity in repeated reasoning.

The practice standards very visibly value reasoning, sense-making, and the use of viable arguments. They are meant to encourage teachers to promote discussion among students of their work. Justifying one's work and making sense of the structure involved are important considerations. Textbooks and resources were adapted to incorporate these practice standards especially at the elementary school level and to a lesser extent at the high school level. Problems that value the *why* and not just the *how* were developed, and intentional work is being done around how mathematics is practiced in schools to make it more engaging.

Canada does not have a unified set of standards, with different provinces adopting their own standards. Here too we see a recent move to adopt standards that speak to more than just mathematical content. In Ontario, for example, the 2020 standards now include a statement on social-emotional learning skills and mathematical processes (see Figure 6.2).

Figure 6.2 • *Ontario Mathematics Standards: Socio-Emotional Learning Skills and Mathematical Processes*

Students will develop social-emotional learning skills and use math processes (for example, problem solving and communicating) across the math curriculum. Students will learn to:

- Make connections between math and everyday life, at home and in the community.
- Recognize mistakes and learn from them.
- Use strategies to be resourceful in working through challenging problems.

As you can see, there is an emphasis on centering mathematics on students' lives by making connections between math, home, and community as opposed to teaching math as a series of disconnected steps devoid of context. There is also language around mistakes as opportunities for learning as well as on incorporating challenging problems into our pedagogies.

Professional organizations such as the National Council of Teachers of Mathematics (NCTM) have long been supportive of nontraditional approaches to the teaching of mathematics as evidenced in their standards documents throughout their history. Most recently they have released the Catalyzing Change initiative meant to ignite change in the way we teach mathematics. They embarked on a comprehensive review of mathematics education at the K–12 level and created a series of professional resources meant to bring instruction away from the traditional school mathematics many of us experienced and toward an active, creative, and rigorous instruction model grounded in critical thinking, communication, and sense-making. We can see from their efforts, as well as from the changes in the standards documents discussed prior, that efforts to move away from passive learning and to focus on procedural over conceptual understanding are underway. Certainly, there are pockets of success throughout Canada and the United States when it comes to the teaching of mathematics in nontraditional ways. However, progress has been slow. Nontraditional approaches are not yet mainstream, and as such, there is still work to be done. Do you see evidence of a move toward nontraditional approaches in your school? How might you encourage this among colleagues?

HOW CAN WE CHALLENGE OUTDATED TEACHING METHODS?

I have already mentioned several of the ways we can challenge traditional views of mathematics and outdated forms of teaching. Using rich open-ended problems, creating space for productive struggle, focusing on conceptual understanding over the teaching of procedures, and actively engaging students in mathematics are some of the ways teachers can move beyond traditional methods of teaching. Next, let's look at a handful of other methods that we as teachers can employ to actively move beyond the ways that school mathematics has traditionally been taught—and encourage our colleagues to do so as well. If you find yourself teaching in a progressive way and incorporating some of these ideas already, consider starting a book club at your school, facilitating a professional development opportunity, or engaging in mentoring so that they can be disseminated more broadly. As other texts exist that flesh out these and related ideas, this is not meant to be an exhaustive list. These suggestions are not lengthy or deep. They are meant to get us thinking about how we might begin to transform our classrooms into spaces where students can grow in understanding of mathematics while developing healthy beliefs about the discipline.

Incorporating History

Mathematics, as all disciplines, has a history. In the case of mathematics, this history is long, diverse, and rich. It spans all corners of the world, involving every civilization to ever exist and a cast of characters so exceptional it is riveting to behold. Yet despite a very rich and colorful history, very often mathematics is taught completely devoid of history. To strip mathematics of the stories that surround it and the mathematicians who developed it is to deprive students of the ability to see mathematics as a truly human endeavor, one that is made real by the many individuals who contribute to its flourishing. It also puts the content at arm's length when we fail to connect with the people and realities that developed the content to begin with. As the author and master mathematical storyteller Sunil Singh (2021) states in his text *Chasing Rabbits: A Curious Guide to a Lifetime of Mathematical Wellness*, "Teaching mathematics without its stories is emotionally empty" (p. 149). In fact, he goes further, arguing that to truly appreciate mathematics and to promote wellness—a concept in opposition to mathematics anxiety—one must routinely get lost in mathematics by "chasing ideas, problems, conundrums, puzzles, and stories of mathematics with unbridled thirst and curiosity" (Singh, 2021, p. 6). Central to this endeavor are stories that bring us into the field and ignite a passion and curiosity for the subject.

Resources: History of Mathematics

Berlinghoff, W. P., & Gouvêa, F. Q. (2020). *Math through the ages: A gentle history for teachers and others* (Expanded 2nd ed., Vol. 32). American Mathematical Society.

Joseph, G. G. (2010). *The CREST of the peacock: Non-European roots of mathematics*. Princeton University Press.

If we are to tell the stories of mathematics, as I believe we must, we need to be able to tell the stories of diverse cultures and peoples. Mathematics has existed in every single civilization that has existed on this Earth. That the stories of non-Western mathematicians have been excluded from school mathematics and the textbooks we rely on is troublesome not only because it erases the contributions of these individuals, but also because it gives students the false belief that mathematics is a strictly Western endeavor. Further, it keeps students from diverse backgrounds from seeing the diversity that exists and has existed in the field, one that they might not only welcome but connect with.

One such story is that of Srinivasa Ramanujan, pictured in Figure 6.3, who was an Indian mathematician during the time of British rule. Born in Madras, India, he was a self-taught mathematician who claimed that mathematical insights came to him from the goddess Namagiri Thayar and stated that equations had no meaning to him unless they expressed thoughts from God.

Figure 6.3 • *Picture of Ramanujan*

SOURCE: Konrad Jacobs/Archives of the Mathematisches Forschungsinstitut Oberwolfach.

His mathematics was too advanced, new, and different to be understood by many, and so for many years he worked in isolation, unable to get others to understand his work. He sent letters to mathematicians explaining his mathematical ideas, and his work was finally accepted by the British mathematician G. H. Hardy. Hardy invited him to live and work at Cambridge University after announcing that he had never seen anything in mathematics like the work of Ramanujan. He became one of the youngest Fellows of the Royal Society and the first Indian to be elected Fellow of Trinity College, Cambridge. His success at Cambridge came with much difficulty as he was away from his family, including his wife, in a country whose customs varied greatly from his own. He later became very ill, ultimately securing a position at a research university in India so he could return home and sadly dying at the age of 32. His story is the subject of the 2015 film *The Man Who Knew Infinity*, with Dev Patel and Jeremy Irons (Brown, 2015). How much more lively, real, and relevant might mathematics feel if we embedded in its teaching the stories of those who contributed to it, complete with their quirky customs and extraordinary lives?

Connecting Math to Student Experiences and Interests

A further consideration with respect to the teaching of mathematics is our ability to connect the content to the lived experiences of our students. I have seen a T-shirt that claims that mathematics is the only place where someone can buy 33 watermelons, and no one questions why. Too often, the contexts used in mathematics problems are so far removed from students' experiences that they are without meaning for them. With mathematics all around us, it seems possible to center student experiences in our teaching of the subject. In this way students can connect to the contexts used and the questions raised, for if the mathematics was taught while valuing the experiences of those from marginalized communities, they would not have such trouble connecting with it, valuing it, and using it. Still, it is the white, middle-class so-called mainstream that is valued in textbooks and other resources. That contexts differ greatly from the lived experiences of the students learning them makes connecting to the material a challenge, which in turn makes learning it a struggle.

In a critical pedagogy course that I taught, students considered how mathematics could be used to explore issues that were of interest to them. They designed projects that would rely on mathematics, which varied greatly from one to the next. One of the students was concerned with the customer service aspect of some of the offices at the college and used survey methods to determine the issues of most concern before presenting these and solutions to the student government, which she ultimately joined. Another used financial mathematics to explore the feasibility of using empty buildings around campus now owned by banks after owners had defaulted on their mortgages as student housing. She researched how proximity to campus improved student outcomes on several measures of student success and then designed a proposal that would benefit both the banks and the students involved, sharing that with various stakeholders. A third looked at the awareness around enrichment programs at the college and used statistics to both understand the problem and test solutions (see Gonzalez, 2018). By exploring their own questions, students were excited about their work. It held meaning for them. Mathematics became a tool that supported their interest and therefore served them.

BEYOND THE TRADITIONAL IN SCHOOL MATHEMATICS

If you want to learn more about how you can transform your classroom into one where students are engaged and active learners, I strongly recommend the following texts. They are among the most excellent current resources we have on the teaching and learning of mathematics. I urge you to read them, share them, and incorporate the ideas therein into your teaching.

Resources for Teaching Mathematics

Boaler, J. (2022). *Mathematical mindsets: Unleashing students' potential through creative mathematics, inspiring messages and innovative teaching.* Jossey-Bass, a Wiley Imprint.

Bush, S. B., Karp, K. S., & Dougherty, B. J. (2020). *The math pact, middle school: Achieving instructional coherence within and across grades.* Corwin.

Dougherty, B. J., Bush, S. B., & Karp, K. S. (2020). *The math pact, high school: Achieving instructional coherence within and across grades.* Corwin.

Karp, K. S., Dougherty, B. J., & Bush, S. B. (2020). *The math pact, elementary: Achieving instructional coherence within and across grades.* Corwin.

Kobett, B. M., & Karp, K. S. (2020). *Strengths-based teaching and learning in mathematics: Five teaching turnarounds for Grades K–6.* Corwin.

Liljedahl, P. (2020). *Building thinking classrooms in mathematics, Grades K–12: 14 teaching practices for enhancing learning.* Corwin.

Matthews, L. E., Jones, S. M., & Parker, Y. A. (2022). *Engaging in culturally relevant mathematics tasks: Fostering hope in the elementary classroom.* Corwin.

Matthews, L. E., Jones, S. M., & Parker, Y. A. (2023). *Engaging in culturally relevant mathematics tasks: Fostering hope in the secondary classroom.* Corwin.

Van de Walle, J. A., Karp, K. S., & Bay-Williams, J. M. (2016). *Elementary and middle school mathematics.* Pearson Education UK.

I wonder how different most people's relationship with mathematics would be if they grew up learning mathematics in classrooms that addressed the realities just described as well as incorporated the suggestions of the resources listed. How much more confident might students be in their abilities, how much more willing might they be to play with mathematics, and how much more likely would they be to derive joy from it? I wonder how much more likely those in society would be to create stories, articles, books, and movies where mathematics is portrayed as a living, exciting, developing subject and those who are engaged in it as diverse, nuanced humans. Might we inspire the next generation of individuals to also excite in mathematics and cringe at the thought of one admitting one is bad at it?

How can you challenge traditional teaching and learning in school mathematics?

- Use open-ended problems.
- Incorporate active learning.
- Encourage productive struggle.
- Bring history into the classroom.
- Center mathematics on students' experiences and histories.

Questions for Reflection

For Teachers

- How can you incorporate active learning into your teaching?
- How can you bring the mathematical practices into your teaching?
- In what ways can you value communication, conceptual understanding, and struggle in your teaching?
- How do you work with parents whose only experiences with the teaching and learning of mathematics might be very traditional?

For Instructional Leaders

- In what ways can you support teachers in moving beyond traditional teaching methods?
- How can you support teachers in meeting the practice standards?
- In what ways do you incorporate history, active learning, communication, and conceptual understanding in your own work with teachers?

For Administrators

- How can you support teachers who wish to move beyond traditional teaching methods?
- In what ways are teachers empowered to choose the materials and resources they need to teach in engaging ways?
- What resources and professional learning opportunities do you provide around nontraditional teaching?
- How do you engage parents and community members, especially around understanding the need for nontraditional teaching?

CHAPTER 7

MATHEMATICS AS GATEKEEPER

In this chapter we will:

- Consider the role that mathematics plays as a gatekeeper to future success.

- Explore the role of standardized testing in mathematics education.

- Review research in the theory of formal discipline.

- Consider alternatives to high-stakes standardized testing as forms of assessment.

- Reflect on how you can challenge the use of mathematics as a gatekeeper to future success and undertake authentic assessment.

THERE'S A MATH TEST FOR THAT

In my first few years of teaching, when my high school students would ask why they needed to learn mathematics and when, if ever, they would use the mathematics we were learning, I would go out of my way to find a connection between the mathematics we were doing and the real world. At the time, I taught algebra and precalculus almost exclusively. There we were in the middle of a lesson on parabolas, and the question would be asked: "Ms. G., no offense or anything, but when am I ever going to use this?" Almost immediately my mind would run through several applications of parabolas to the real world and rattle off an argument that one day they might be designing a bridge or a reflective lens for a telescope for which the math we were learning would be incredibly useful. When they insisted that never would they be designing a bridge or parts for a telescope, I would counter that one never knows what the future will bring. My argument fell flat onto the floor with a huge thud every time. While I found an application of parabolas in the real world, it wasn't *their* world or *their* perceived future world that my examples were coming from. The examples in the texts that we relied on and the examples that most easily came to my mind often centered on experiences and realities that were far removed from my students. The material wasn't of interest, and—if I am being honest with myself—they were right to challenge my argument.

As years passed, I began to see all too clearly one of the biggest ways in which the content I was teaching did impact the lives of my students. It was a gatekeeper. Reaching one's goals often depends on performance on a mathematics exam (or multiple exams). If one wants to graduate from high school, there is a mathematics exam to pass. If one wants to go to college, there is a mathematics exam to take. If one wants to obtain a civil service job and be promoted in that job, there is at least one mathematics exam to take. Law school, medical school, and schools of education require that students pass exams with mathematical content to be admitted. Graduate school, often regardless of the actual field of study, also requires an exam with mathematics content for entry. In most of these cases, regardless of the area of study or profession, the mathematics tested is basic arithmetic, algebra, and logic. With time, I learned to be honest with my students and spoke to them about the fact that there would be a mathematics exam (if not multiple exams) between them and their dreams, while also working to center the content on their experiences in a way that better connected the material to their lives. I often joke that if dancing were used in a similar gatekeeper manner, I would likely not have the position I do today. Fortunately for me, the discipline that is valued is the one I love and excel at, but this is not the case for many. It is time for us to be honest about the ways in which mathematics is used in our society to filter students out of opportunities. This is done through an overreliance on standardized high-stakes mathematics tests, but need it be so? To be clear, in this chapter I am focused on summative assessments, benchmark assessments, and high-stakes standardized exams as opposed to formative assessments.

ALGEBRA AS A CIVIL RIGHT

As exams focused on algebra are used throughout our society for entry into college and career opportunities, the teaching of algebra is critical. Moses and Cobb (2001), viewing the role that algebra plays, liken the importance of access to high-quality mathematics instruction in algebra for students to Black people obtaining the right to vote. Without a solid grounding in algebra, they believe that students will be unable to engage fully in society and, therefore, be denied opportunities that may lead to their future success. The right to algebra is a matter of social justice.

> **The right to algebra is a matter of social justice.**

Robert Moses was an educator and civil rights leader who founded *The Algebra Project*, an organization that aims to help students from low-income backgrounds and students of color have access to high-quality mathematics instruction so that they are college ready upon graduating from high school. The organization engages in efforts to increase mathematical literacy by providing curricular materials, teacher training, professional development, and activities for schools to engage the broader community.

In our society as it presently exists, knowledge of algebra opens doors. It also allows for further study of the mathematics that is presently valued in society, which in turn opens more doors. To study the role of mathematics as a gatekeeper to future success, Daniel Douglas and Paul Attewell (2017) used data from the Education Longitudinal Study of 2002, which concluded in 2012. The data included information on 15,000 U.S. students followed for 10 years beginning in their second year of high school. What the researchers found is that high school mathematics—as measured by test scores and/or coursework completed—was associated with several higher education milestones including attending a four-year college, attending a more selective college, and completing a bachelor's degree. This was true even after controlling for other variables such as family background, general academic performance, and motivation, as well as other forms of social advantage. Taking calculus in high school led to a 16-percentage-point increase in the probability of attending a four-year college as compared to students who had taken only Algebra 2. The probability of attending a selective college went up 18 percentage points for students who took calculus in high school as compared to students who had taken only Algebra 2.

There is no doubt that algebra because of our reliance on mathematics testing as a filter is a necessary skill for acquiring access to opportunities. I wonder, however, if there are better ways to assess who can excel and who should be given those opportunities. I wonder too, were we to rely less on algebra as a gatekeeper, if we could free up some space in the curriculum for other perfectly valid mathematics that currently gets neglected.

MATHEMATICS AND COLLEGE COMPLETION

In college, introductory-level mathematics courses continue to be the most significant barrier to students' completion of a degree, as virtually all majors require that students pass a mathematics course. Many struggle with this course. In general, only 5% of college students in the United States pass a gateway mathematics course in their first two years, yet without this course they cannot graduate from college (Complete College America, 2021). In some cases, this mathematics course might be a prerequisite to further courses needed for their degree. This is one of the reasons why there are 36 million Americans with some college credit and no degree. Roughly 4 in 10 college students who began their studies in 2012 did not earn a degree six years later (Shapiro et al., 2018).

The numbers are more troubling when we consider Black and Latinx students (National Center for Education Statistics, 2019) as can be seen in Figure 7.1.

Figure 7.1 • *College Completion Rates in the United States*

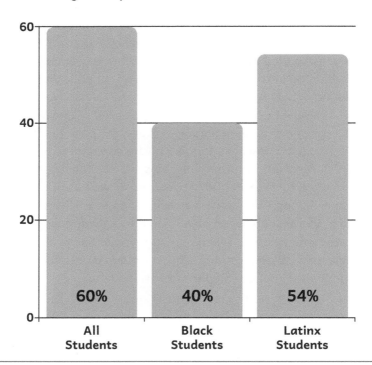

While there are many reasons for this, one big stumbling block is mathematics. A third of students cite academic performance as the reason why they leave school, with as many as a quarter of all students being placed into remedial courses upon entry. That means that students are placed into classes where they pay tuition but do not earn credit, adding to the time needed to complete a degree as well as the amount of student debt they accumulate. It also makes them much less likely to complete a degree at all.

EXPLORING THE THEORY OF FORMAL DISCIPLINE

That mathematics is a gatekeeper to future success is well established, but why mathematics? The argument usually made for using mathematics as an entrance requirement is that math, most especially algebra, is a predictor of future success. However, is mathematics a good *predictor* of future academic and career success because it is inherently a good *measure*? Of course, mathematics predicts future success in that without passing mathematics exams one cannot proceed in one's academic, and often one's career, trajectory. This does not necessarily mean that the content we are testing is a good predictor of the skills and abilities needed to succeed in life. It is, however, very good at predicting that those who excel at it will have an easier time accessing and succeeding in a system of schooling that constantly tests students in mathematics (most especially algebra) to progress. It is a self-fulfilling prophecy because of the way the system is designed.

The idea that the study of mathematics develops critical thinking is part of the theory of formal discipline. The theory posits that mathematics can lead to improved reasoning and skills that transfer to other fields—or, put simply, that mathematical study makes one a better thinker in general. The theory of formal discipline has been supported throughout history and used as a reason to include mathematics in the curriculum. To test this theory, Inglis and Attridge (2017) engaged students in different reasoning tests. Their work was based in England, where students are not required to study mathematics beyond the age of 14. There, students can participate in a one- or two-year sequence of courses in particular areas known as AS (one-year) or A-levels after completing compulsory mathematics education if they wish to prepare for university study. Some students elect to take a rigorous one- or two-year course of study in mathematics while others take the one or two years in other subjects. This allowed the researchers to test students whose additional pre-university studies focused on mathematics and to compare those students to others whose focus did not include additional mathematics training. Students were invited to complete a series of questions testing reasoning skills that were modeled after some previously developed questions used in well-known studies in the past. It was found that the students who chose to study math did considerably better than the non-mathematics group across several different types of logic problems. The differences were more pronounced when the logic problem was abstract in nature rather than when it was couched in a social context, but they existed across all the categories of problems used. Thus, if we are to believe that the study of mathematics improves logical reasoning skills, it seems to do so in more abstract cases.

While it might seem that their research supports the idea that studying mathematics boosts logical reasoning skills, a different interpretation would be that the students who choose mathematics are more skilled in logical reasoning to begin with. Therefore, it is not the study of mathematics that makes them better logical thinkers, per se. It is, rather, that better logical thinkers tend to study mathematics at higher rates than those whose reasoning skills are less developed. With that in mind, Inglis and Attridge (2017) undertook another series of experiments. This time, they tested students who were studying mathematics A-levels but did so at two different times: close to the beginning of their studies and again closer to the end of their first year of engagement in this noncompulsory mathematics program. It should be stated that the curriculum for this program did not include a study of logic and of problems like the ones they were being asked as part of the research study. Here, the researchers found that the students' logical reasoning did improve over time and with exposure. The students' answers on the second round of logical reasoning tests were better, and markedly so, than they had been the first time around. There was overall improvement in the reasoning abilities of the mathematics students as they proceeded throughout their studies.

DOES REASONING ON A TEST RELATE TO REASONING IN REAL LIFE?

One might wonder if answering a few questions on a research study impacts one's ability to make good decisions and use proper logical reasoning in the real world. What if the improvements we saw in reasoning skills are limited to the types of problems the students were being asked? What if mathematical study improves logical reasoning to some extent but that improvement is not transferred into the real world? Here we rely on the work of Bruine de Bruin and colleagues (2007). They explored whether individuals showed good use of logical reasoning skills in their everyday lives by asking how often participants had been locked out of their homes; whether they had missed a flight, taken the wrong train, or bounced a check; and how often they threw out expired food. A good number of scenarios were presented where those who could problem solve and use reasoning in the real world would be more likely to perform better—to make their flight and be less likely to throw out expired food. These same individuals were then asked to complete a series of logical reasoning tasks in a more traditional laboratory setting. It was found that individuals who did better on the in-lab logical reasoning questions also did better on real-world problem-solving and reasoning tasks. Therefore, there is every reason to believe that improvements brought on from the study of mathematics indeed translate into real-world improvements in one's reasoning and problem-solving abilities.

As a result of their research, Inglis and Attridge (2017) suggested that the study of mathematics improves logical reasoning and leads to changes in how students are able to think through such problems, most specifically those that are more abstract, though they acknowledge that the improvements seen are not as dramatic as one might hope for and expect. Therefore, it seems reasonable to ask whether a shift from such an algebra-heavy curriculum to one that offers a sufficiently solid grounding in algebra while increasing exposure to other areas of mathematics such as logic, discrete mathematics, and financial mathematics might be both more appealing to students and more useful to them as they go out into the world.

RETHINKING THE ROLE OF MATHEMATICS TESTING

It might also make sense to lessen the gatekeeper role of mathematics by not relying so heavily on it as a prerequisite to college entry, entry into graduate programs, and entry into civil service jobs. Can we not adjust these assessments to lessen the weight given to mathematics in favor of work products or other forms of assessment that capture the whole person and their abilities? At present we expect that students who do not have the prerequisite mathematics (typically algebra) skills will not excel in college and beyond, but does this expectation keep a studio art major or a film major from pursuing their dreams while allowing those who favor the sciences more of a shot? It seems to. If we adjust the assessments to allow students more flexibility in accessing college education and then adjust the courses of study to broaden what mathematics we expose students to, they might find that they can indeed excel.

The Misuse of Standardized Exams

The use of high-stakes standardized exams is extremely prevalent in the United States, though much less so in Canada. In the United States students take an average of ten standardized exams a year in Grades 3–8 (Lazarín, 2014), whereas in Canada the number of total exams in Grades 3–12 is roughly five (Després et al., 2014). Most standardized tests in Canada do not count toward a student's grade, and the uses of these exams are not as broad as they are in the United States. That is, they are often not high stakes. Still, both countries have seen an increase in the use of these exams in the recent past. Further, these exams are being used for a myriad of purposes from comparing student achievement to assessing teacher performance, determining school success, and informing funding formulas. These constitute a misuse of the exams with "the primary issue being the expectation that tests work as a definitive sorting tool for children, teachers, schools, districts and jurisdictions—a task for which they were not originally designed and one which they do not have the capacity to adequately perform" (Kempf, 2016, p. 41).

Negative Effects of Standardized Testing

Further complicating matters is the fact that when questions are presented in a context, this context is rooted in the experiences of white students from middle-class families, as is the case with standardized exams in general. This disadvantages Black and Latinx students, who do not see their lived experiences reflected in the assessments used. Additionally, high-stakes exams affect instruction in schools that serve students of color from lower-income families. Often, in these schools such a strong emphasis is placed on these exams that the school may rely on more traditional curricula that focus on test prep rather than more progressive pedagogy (Faulkner et al., 2019). More progressive forms of pedagogy that center instruction on students' experiences, value mixed-ability groupings, and adopt a constructivist model have been shown to be more effective with respect to engaging students as well as improving understanding, especially among those from marginalized communities. However, these approaches tend to be used less in contexts where the exams carry a lot of weight and the school in question is underperforming. With a focus on the study of the impact of high-stakes standardized testing on pedagogical practices in the United States and Canada, Arlo Kempf is an assistant professor of equity education and teacher development in the Department of Curriculum, Teaching & Learning at the University of Toronto. Kempf (2016) quite literally wrote the book on standardized testing in the United States and Canada. In his text,

The Pedagogy of Standardized Testing: The Radical Effects of Educational Standardization in the US and Canada, Kempf found standardized testing narrowed instructional content, with well over half of the respondents agreeing or strongly agreeing with the statement, "As a result of standardized testing I cover a narrower range of topics than I would otherwise." Teachers were also asked about the use of diverse assessment and instructional approaches, and here too standardized testing has an effect, with teachers using less innovative methods as a result. When testing drives instruction, "rote learning and what many call drill and kill replace project-based exploratory approaches, to the detriment of students who benefit from varied teaching and learning strategies" (Kempf, 2016, p. 56).

> When testing drives instruction, "rote learning and what many call drill and kill replace project-based exploratory approaches, to the detriment of students who benefit from varied teaching and learning strategies" (Kempf, 2016, p. 56).

See Figures 7.2 and 7.3 for a summary of Kempf's findings with respect to these two areas. It is sadly ironic that the pedagogies that benefit students tend to be dismissed in settings where the pressure to perform well on standardized exams is critical to the school in question, but this is exactly what occurs.

Figure 7.2 • *Impact of Standardized Testing on the Range of Topics Teachers Cover*

As a result of standardized testing, I cover a narrower range of topics than I would otherwise:

	Ontario English	Ontario French	Illinois
Strongly agree	19.1%	45.6%	27.2%
Agree	38.4%	30.8%	35.5%
No opinion	19.9%	9.5%	14.9%
Disagree	15.9%	10.3%	17.4%
Strongly disagree	6.6%	3.8%	5.1%

Figure 7.3 • *Effects of Standardized Testing on Teachers' Ability to Use Diverse Assessment and Instructional Approaches*

Standardized testing impacts my ability to use diverse assessment and instructional approaches (such as differentiated instruction and multiple intelligences) as follows:

	Ontario English	Ontario French	Illinois
Makes it much easier	0.6%	1.0%	1.0%
Makes it easier	2.1%	3.4%	5.6%
Uncertain	26.4%	29.9%	23.5%
Makes it more difficult	42.3%	37.4%	43.6%
Makes it much more difficult	28.7%	28.4%	26.3%

A further drawback of constant high-stakes testing is the stress that it places on students, teachers, and families. Boaler (2015b) notes that mathematics teachers test more often than teachers of any other subject because they often mistakenly believe that mathematics is about performance. Further, in various studies, students who received regular feedback on their work but no grades were more willing to learn and were more positive about learning compared to those who received grades alone or grades with comments (Elawar & Corno, 1985; Pulfrey et al., 2011). In fact, it is recommended that teachers limit the amount of grading they do in favor of instructive comments. In his book *Grading for Equity*, Joe Feldman (2019) explains how traditional grading practices—including the 100-point scale, using a grade of zero for missed assignments, and grading homework—demotivate students with respect to learning and lead to grading that is both inequitable and an inaccurate measure of their ability with respect to the content. Partly, this is because grades are an extrinsic motivation and giving grades for every assignment as well as for participation leads to a culture where points are valued over learning. Instead, Feldman suggests practices such as using a 4-point scale, allowing for late assignments, and permitting students to redo assignments to suggest that what matters is learning the content and mastery over time. He also suggests weighing assignments at the end of the term more heavily than those that occur when students are just being introduced to material, as well as not grading homework and not giving a grade for participation or effort. The idea is to foster conditions that limit bias, increase accuracy, and motivate students to learn the material, not merely try to get good grades and earn points however they can. By addressing grading practices, we can center learning as the core activity rather than the acquisition of points.

Resources: Assessment

Dueck, M. (2014). *Grading smarter, not harder: Assessment strategies that motivate kids and help them learn.* ASCD.

Dueck, M. (2021). *Giving students a say: Smarter assessment practices to empower and engage.* ASCD.

Feldman, J. (2019). *Grading for equity: What it is, why it matters, and how it can transform schools and classrooms.* Corwin.

REPLACING STANDARDIZED EXAMS WITH AUTHENTIC ASSESSMENT

What if, instead, we replaced these high-stakes assessments with more nuanced and holistic assessments that better reflect what students know and required them to put that knowledge into action? A project that is designed and carried out by the student, that is done over a period of several months, and that relies on mathematics is a viable option. Students can, with guidance, develop a proposal for their project. They may design a roller coaster, use their knowledge of mathematics to track the spread of disease, or use math to study social, historical, or natural realities. In this way they would be putting the mathematics they learn to good use in a capstone project that relies on research, creativity, ingenuity, and mathematics. Further, they could be combining subjects, using interdisciplinary knowledge to both carry out the project and present it to a review board of teachers and fellow students. Understanding material enough to use it and communicate it in both written and oral form is a crucial life skill whether we are talking about mathematical knowledge or knowledge of other fields. More than an exam, such projects give students a sense of what it is like to use mathematics in the real world. Projects are organized into graduation portfolios, giving students a work product that can be used in interviews with future employers as well as written about in application essays to college.

To assess student understanding in the classroom:

- Consider replacing typical exam questions with alternative assessments for gauging student understanding.

- Students can create *all about me* posters for numbers and concepts to demonstrate their knowledge about these. For example, the number 3 is prime, odd, a factor of 12, and so on. Let students get creative with the poster and allow them to decorate it if they wish.

- To demonstrate an understanding of reflections and rotations, have students design butterfly wings (for line symmetry) or decorate the petals on a flower (for rotational symmetry).

- To show they know how to graph functions, have students create a picture using functions and provide the equations and their domains. Students can even take each other's equations and re-create their classmates' pictures. This can be enhanced by the use of technology such as *Desmos*®, a free suite of math tools that includes a graphing calculator and is used by over 75 million people each year (www.desmos.com).

Lastly, relying on capstone projects puts the task of assessment squarely on the teachers who have worked with these students all year long and who know both the students and their abilities well. Presently, summative assessment is controlled by companies on which we spend billions of dollars to develop not just the tests themselves but also curriculum aligned to the tests and test-prep materials while the true educational experts—teachers—are relegated to the sidelines. Further, the money being spent on testing can be redirected to the schools themselves, providing much-needed supplies, allowing for the hiring of more teachers to reduce class size, and building more schools to ease overcrowding. The money can be used to create programs that enhance educational opportunities and provide enrichment for students, especially those at schools whose students are traditionally underserved.

Some erroneously believe that standardized testing is an inevitable piece of the educational puzzle—that is, that to evaluate students requires the level of standardization such exams bring. However, academic portfolios provide a more robust assessment, and their use and effectiveness have already been proven. This model is presently being used by over 35 schools in New York City that have obtained a waiver from the state for Regents Exams. These schools participate in the New York Performance Standards Consortium, founded two decades ago, and use nonstandardized assessments for their students.

In this model, assessments are not imposed top down on curriculum but rather grow out of the curriculum. Teachers are intimately involved both in designing curricula that draw on student interest, valuing student voice and choice, and in developing assessments that measure their growth and achievement.

Questions for Reflection

For Teachers

- How can you design authentic assessment tasks that move beyond typical exams?

- In what ways can you engage others in discussions around the role of mathematics as a gatekeeper?

- What nontraditional assessments can you develop in your classroom?

- How can you work with parents to ensure they are aware of and prepared for the role that mathematics plays as a gatekeeper for future success?

For Instructional Leaders

- In what ways can you support teachers in moving beyond traditional assessment methods?

- How can you advocate for more robust and holistic assessments in the school settings in which you work?

- In what ways do you incorporate holistic and nontraditional assessments in your own work with teachers?

For Administrators

- How might your school adopt nontraditional assessment methods?

- In what ways can you support parents who wish to opt out of standardized exams?

- How are the methods you use to evaluate teachers and staff holistic?

- How do you engage parents and community members, especially around understanding the role of mathematics as a gatekeeper to future success?

- Can your school opt out of standardized testing to some degree?

- How might the placement policies in your school allow all students the opportunity to take rigorous mathematics courses?

- What support do you offer families around standardized testing?

CHAPTER 8

ACHIEVEMENT GAPS OR OPPORTUNITY GAPS?

In this chapter we will:

- Reframe achievement gaps as opportunity gaps.

- Consider what a lack of instructional and personnel resources means to mathematics teaching.

- Explore how the move to remote learning due to the COVID-19 pandemic is an example of an opportunity gap.

- Consider how you can use your privilege and position to challenge the existence of opportunity gaps in math and beyond, across educational settings.

You may notice that Chapters 8 and 9 shift a bit to examine some of the broader systemic issues around education. While certain steps may be taken to start dismantling the *bad at math* stereotypes and tropes that reinforce inequitable practices and policies at the classroom and building levels, they are only Band-Aid solutions if we don't also address the underlying systems in which our buildings and classrooms function. These discussions reach well beyond the subject of math or the walls of the math classroom but are nevertheless important to have as part of an effort for full systemic change to mathematical inclusion and access for all students.

THE HARD TRUTH OF THE HAVES AND THE HAVE-NOTS IN EDUCATION

I began my career as a high school math teacher. I taught in a large, underperforming New York City public school for roughly seven years. It was there that I saw the promise of education smack up against the injustices of our society. Our students were not any less capable than those at schools that served students from more affluent backgrounds, but the resources and opportunities afforded to those other students were not regularly afforded to ours. Our students had less access to Advanced Placement (AP) courses, enrichment opportunities, robust after-school programs, and elective courses. Our facilities were not as modern or well kept, and our financial and instructional resources not as plentiful. Further, and sadly, our students were not expected to do as well.

Several years ago, I had the opportunity to tour a well-known elite private school in New York City. The school serves students in Grades Pre-K through 12. The differences between this school and the one I taught at were stark. As I walked in, I was struck by how well lit the hallways were and by the large windows that let in a beautiful amount of sunlight and were devoid of the bars I was accustomed to in my classroom. There was air-conditioning throughout the school. The classrooms were larger and brighter, and each one had twice the number of teachers and half the number of students. There was a swimming pool, an indoor track, indoor tennis courts, and a basketball court. There was a robotics lab, a STEM lab, and various art studios. All students were taught how to swim and taught a language other than English from the time they were toddlers. They had a choice of five world languages to study. In music class, students were taught to play various instruments, and with each one they learned, they had their own that they could take home to practice with. The school had a speaker series where members of government, scientists, and those in the arts came to the school to speak to students. The woman leading the tour shared with me a copy of a book of poetry that the third-grade class had published, and she pointed out the beautiful artwork made by one of the older grades along one wall. Tuition was over $40,000 a year, an amount well above what the families in my school could afford and, in some cases, greater than these families' entire income for the year. The stark differences exist not only between private and public schools, but among public schools as well. It is not uncommon to see features similar to this private school in public schools in affluent neighborhoods. I encourage you, if you can find the time, to take tours of schools that serve students from different socioeconomic backgrounds. It is eye-opening.

While I knew our students were not being served well, it was incredibly clear just how large the difference in resources was when it was in front of me in the form of a large, multilane swimming pool in a well-lit, well-funded, well-connected, well-resourced school. It is beyond troubling to me to know that some students have so much while others have so little. I wonder how much more the students in my school could have achieved and how much they could have benefited from an environment with the resources and support of that private school. It is also an insult to such students that they learn in a space that is not as well cared for, not as modern, and not as well supplied. What do broken-down buildings and a lack of resources tell the students who must each day work in these conditions about the value that we as a society place on them and their education?

> **What do broken-down buildings and a lack of resources tell the students who must each day work in these conditions about the value that we as a society place on them and their education?**

It tells them simply that who they are doesn't matter—that they do not matter enough to invest in their environment, their education, and their lives.

It was not just that the physical spaces were more modern and well kept, that the resources were more plentiful, and that the social connections were more influential, but there were other differences as well. Teacher turnover was high at the school where I taught. After only seven years, I was among the most senior teachers in a department of roughly 20 math teachers. High teacher turnover and high numbers of inexperienced teachers are common realities in schools that serve students from lower-income families and students of color. Teacher turnover rates are 50% higher for teachers in schools serving students from lower-income families and 70% higher for teachers in schools serving the largest concentrations of students of color (Carver-Thomas & Darling-Hammond, 2017). Another common occurrence is classes taught by teachers who are out of license. That is, it is common for teachers who are licensed to teach one subject to be teaching a different subject altogether.

In my first year of teaching, I was assigned a special education class where students had various learning and emotional needs that I was not trained to meet. I did not have a license in special education or any training in how to work with students with disabilities. I gave it my all each day, knowing I was failing them miserably. Rather than hire a special education teacher, the school assigned the class to me, as I needed one more course to have a full schedule. I was woefully unprepared, and the students in that class suffered as a result. Filling a teacher's schedule, with a class that the teacher is not prepared to teach, is an immediate-term cost-saving measure. It costs less to pay one teacher to teach five classes (a full load)—where one of these classes is outside their license area—than to pay for an additional teacher to teach the one class. But what is the long-term cost, and for whom? It is the students who ultimately pay.

Similarly, on October 3 of a particular school year, I found myself covering a mathematics class for what I assumed was a teacher who was out sick. The students were recent immigrants to the United States, all of whom spoke Spanish, and the course was a bilingual mathematics class. I am fluent in Spanish and so spoke to the students about what they were doing in the class, but they told me that they didn't do much. I asked who their teacher was, and they told me that they did not have one. I found it hard to believe that by October 3 there would exist a class without a teacher, so I asked them who their science teacher was, and again they said they did not have one. "What about your history teacher?" "We don't have one." When I looked at one student's program card, a blue card that lists every class a student has in their schedule, I was shocked at what it revealed. The card listed a class for every period except lunch, but in place of a teacher, the card said TBD for every single class. "So, what do you do all day?" I asked with concern. I was told that they had a series of substitutes all day and that these teachers changed from day to day. Sometimes the teacher would try to teach them something related to their field, as I had done, but often the person spoke no Spanish and could not really communicate with them. On most days they were allowed to pass the time in conversation with each other. It seems that the school had been assigned bilingual students that it was not expecting. Sadly, students who had not chosen this as one of their picks for high school were assigned to the school regardless. As we were a low-performing school, not everyone was there by choice. True school choice does not exist in a setting where the number of seats in desirable schools is less than the demand for them.

> True school choice does not exist in a setting
> where the number of seats in desirable schools
> is less than the demand for them.

As a result, we often had students who didn't want to be at our school, and large numbers of students who were placed with us after the school year started and who were tragically and inhumanely referred to as *over-the-counter kids.* I hesitate to even write that phrase here as no one should refer to a child in this manner, but I include it because it speaks to the ways in which these students are not being respected or served by the system. Thus, it turned out that there were students who needed bilingual classes, but the school had not yet hired teachers to teach them. The school accepted my offer to teach the class, in addition to the classes I was already assigned, until a suitable teacher was found.

I began teaching the class. While having a mathematics teacher, and one who speaks their language, is better than not having a teacher at all, my teaching the course was grossly unfair to these students. I had no training in bilingual education. I had never even taken a mathematics course in Spanish, and while I knew the mathematics, I did not always know the mathematical vocabulary in Spanish. I, again, did the best I could, and together we made it through the material, learning from each other. I was thankful to have some Spanish-language texts that the school had acquired. I taught the class from the beginning of October until February. In all that time, a suitable teacher was not found, though I am unaware of what efforts were made to find one. I was then replaced by a history teacher who was fluent in French. She was clearly not a suitable teacher, given the class content and students involved. Thus, the students' right to high-quality education in mathematics continued to be violated. It seems very hard to imagine that this sort of injustice would have occurred in a school serving students in more affluent areas or one where the population of students was predominantly white.

For those looking for a text focused specifically on English-language learners, I highly recommend *Teaching Math to Multilingual Students, Grades K–8: Positioning English Learners for Success*, written by Katherine Chval and her colleagues (2021). In that text the authors talk about *positioning*, a term taken from social psychology that refers to the rights and responsibilities students are expected and allowed to carry out in social interactions within a classroom. How students are positioned in the classroom and in the school can affect their social, communicative, and academic success. The authors specifically address the positioning of multilingual students in mathematics classes, though there are lessons in this text for mathematics teachers of all students.

Before moving on, I want to note that there were some amazingly wonderful teachers and administrators at the high school where I worked. These were people who truly cared about the students and put their all into the work they did. There were definite pockets of success there, including an International Baccalaureate program, which was the only one at a public school in New York City at the time, for example. There were also some of the finest faculty, staff, and students I have encountered. I do not share the struggles we faced to disparage their efforts. Rather, I want to highlight the fact that we were working within a school system that was and is fundamentally unfair—one that could not be completely undone by the efforts of a few, no matter how devoted and resourceful they were and continue to be.

How does your school compare to others in the city, state, or province in which you teach? What resources are available to you and your students, and how are these the same as and different from other schools in your area? What resources are lacking, and how does this compare to other schools you are aware of? The differences in schooling experiences of some students as compared to others lead to opportunity gaps. That is, we can trace differences in student outcomes in large part to the differences in the opportunities they have. Let's explore this concept and how it differs from the more commonly used achievement gap rhetoric.

REFRAMING ACHIEVEMENT GAPS AS OPPORTUNITY GAPS

The concept of an achievement gap is prevalent in research literature published in academic journals as well as in the mainstream media. Students of color have been and continue to be compared to white students and their peers who are more affluent in other schools with respect to performance. The narrative that has been put forth highlights the so-called achievement gaps between students of color and their white counterparts. This narrative is driven in part by the large amount of data that we collect through standardized testing such as the National Assessment of Educational Progress (NAEP), the Program for International Student Assessment (PISA), and state standardized tests, and by the constant comparisons being made with such data. The purported goals of such research are to understand the different achievement levels of various groups so that an effort can be made to minimize the gaps. With concern around these achievement gaps, we enact policies to support those who struggle in this discipline. Many of these programs focus on providing tutoring, remediation, mentorship, changes in pedagogy, and curriculum with the aim of closing these achievement gaps, but the gaps continue to persist, though in some cases they narrow slightly. One should question whether these efforts are reaching their intended goal or if what we are measuring is less a matter of the achievement of various groups and individuals and more a result of the opportunities afforded them.

In her article "A Gap-Gazing Fetish in Mathematics Education? Problematizing Research on the Achievement Gap," Gutiérrez (2008) writes:

> At their most extreme, achievement-gap studies offer little more than a static picture of inequities in schools. Because these studies rely primarily upon one-time responses from teachers and students, they can capture neither the history nor the context of learning that has produced such outcomes. And, whereas researchers can highlight the variables most closely associated with the gap (e.g., income, family, background), those variables are often not reasonable levers for change in the mathematics education community. Moreover, the cross-sectional nature of most achievement-gap data analysis means that they fail to capture student gains or mobility. . . . Regardless of the form of the data, the theoretical lens used to view achievement gap is what supports deficit thinking and negative narratives of students of color and working-class students. (pp. 358–359)

This focus on achievement gaps is hurting our most vulnerable students who are constantly being compared to the so-called mainstream. Can we not understand their learning without comparing them to historically dominant groups? Can we not assess strengths without the performance of white students being the standard by which others are measured? Not only that, it promotes the wrong narrative. We are looking at the wrong end of things. We are putting the focus and fault on the children themselves when we should be looking at the system that is supposed to serve them. We are focused on the outcomes when we should be looking at what leads to such outcomes. We should be focused on the opportunities, or lack thereof, that students have throughout their academic careers. If we focus on ensuring equity with respect to opportunities, I firmly believe that the outcomes will take care of themselves—that is, that the so-called achievement gaps exist because of how the school system is set up and the incredibly differential opportunities therein. They are not a reflection of the natural ability of the students before us but rather a result of the system in which these students are educated and the society in which they are raised. If we focus not on the results but on the systems, educational and otherwise, what we see are opportunity gaps, not achievement gaps. We need to pay attention, then, to the inputs, the quality of education, and the social and financial opportunities being afforded those who traditionally have been underserved.

WHERE ARE THE OPPORTUNITY GAPS, AND WHY DO THEY PERSIST?

In his groundbreaking book *The Shame of the Nation*, Jonathan Kozol (2005) recounts visiting approximately 60 schools in 30 districts in 11 different U.S. states. He details the conditions he found at schools that serve students in lower-income areas where Black and Latinx students are traditionally overrepresented, including the school where he began his own teaching career. He writes of the deplorable conditions where students were being taught in schools with decaying facilities and gross overcrowding in spaces not meant for learning. He writes also about the segregation by race that exists in school districts across the country that, although not sanctioned by law, persists, and about the repercussions of a system that he terms "apartheid schooling in America." The conditions he describes constitute a very real and very challenging reality grounded in racism and classism that continues to persist to this day and has seeped into every aspect of our social systems, education included.

Opportunity gaps exist and persist at the systemic level. They have to do with inequities in

- well-trained personnel and manageable class sizes;
- course placement;
- access to technology and rich instructional resources;
- community supports, such as access to nutritious food and safe homes; and
- fragmented policies and budgets.

None of these inequities are new, yet in the spring of 2020, the onset and consequences of the COVID-19 pandemic highlighted in a very stark way the inequities present in our society, both educational and otherwise—inequities that will remain unless there is careful and intentional work done around these issues at a societal level. Let's look at each of these issues in more depth, including how the COVID-19 response unveiled and exacerbated some of them.

Inequities in Well-Trained Personnel and Manageable Class Sizes

Having enough licensed, well-trained, properly supported, subject-matter-confident mathematics teachers, special education specialists, and leaders is key if we are to support students, especially those most vulnerable. Due to many factors, including how the profession is viewed and the lack of support and resources afforded them, more teachers are leaving the profession, and fewer teachers are entering it. Further, districts with more than 75% of their population made up of students of color have double the number of teachers resigning as those districts where 90% or more of the students are white (Stanford, 2022). As such, we find ourselves in a situation where some schools are lacking the human capital needed to adequately supply their students' mathematical learning. A lack of access to math coaches and math specialists can also make teaching mathematics effectively difficult, especially for beginning teachers or those without a solid grounding in mathematics. Teachers need mentors and strong mathematics education leaders and coaches to work shoulder-to-shoulder within every building.

Similarly, having adequate time for planning and working with other teachers is another consideration. Opportunities to work with and learn from other teachers can be extremely beneficial, but often time is not dedicated for such work. When teachers have little time to plan, and work in isolation, they tend not to be as effective and, one might add, not as happy either. Happy teachers, as happy people in all professions, are better able to flourish. The same, of course, is true of students.

Another benefit of having an adequate number of well-prepared mathematics teachers and leaders would be the ability to reduce class size. Dee and West (2011) used survey data in middle schools to show that lower class size leads to an increase in noncognitive skills such as more positive reactions to teachers and classmates as well as higher satisfaction with school and increased motivation. It isn't just noncognitive skills that improve from smaller class sizes. Using Tennessee's Student Teacher Achievement Ratio (STAR) project, Konstantopoulos and Chung (2009) found that young students of all abilities benefit academically from being in small classes. Their work is in line with that of Bosworth (2014), who found similar results among fourth- and fifth-grade students in North Carolina. Lastly, Vasquez Heilig and colleagues (2010) conducted research in urban schools in Texas that serve a high percentage of Latinx students and found that "the most powerful predictor of changes in reading and math in all models was decreasing the student–teacher ratio" (p. 52).

Yet despite the research, high class size persists, particularly when faced with a growing teacher shortage. In Toronto, while secondary school classes are capped at 23, an increase in the school-age population is leading to overcrowding, with some schools reporting class sizes as high as 40

(Tsekouras, 2021). In fact, even though the student-to-teacher ratio in Canada is roughly 17 students per teacher, overcrowding is becoming more of a problem in certain urban areas that are seeing an increase in population such as Toronto and Vancouver. However, urban areas aren't the only ones seeing a surge in students. In 2018, the Calgary Board of Education released a report that up to 40 schools in suburban areas were operating at over 100% capacity as well (Ferguson, 2018).

In the United States, the San Joaquin Valley in California has seen a rapid doubling of population due to access to affordable housing as compared with nearby areas (Gallegos, 2022). Young families that have been priced out of nearby areas are flooding into this area, which has put a strain on schools. Class sizes have ballooned. The enrollment in the Sanger Unified School District, for example, grew 58.1% between 2000 and 2019 and another 4.5% in the three years following. As a result, portable classrooms are being placed over basketball courts and athletic fields. What do these conditions do to the teaching of mathematics? How can students do their best work when they find themselves in overcrowded portable classrooms and are denied opportunities for athletics as a result?

Hiring/retaining an adequate number of qualified teachers and lowering class size would go a long way to ensuring that students get the individual attention they need to be more successful. The class-size national average for public schools in the United States is 25 students, while in private schools the average is 19 students per class. Private schools have a better student-to-teacher ratio (approximately 12:1) as compared to public schools (approximately 16:1) across the country (Wang et al., 2019). One way to lower class size specific to mathematics education is to hire additional mathematics specialists. This is especially key at the elementary level, where teacher preparation programs do not always provide a thorough grounding in mathematics and mathematics methods. Many teachers take one mathematics methods course as part of their training and may, without further support, fall back into the traditional ways of teaching they saw in their own schooling.

One argument against hiring more mathematics educators and lowering class size is cost. Not only does the hiring of personnel cost more money, but having more teachers and smaller classes also means finding more classroom space, which may mean building more schools. This argument fails to highlight the reality that budgets, while limited, are a matter of priorities. Investing in schools that serve children from poorer areas and in schools that serve predominantly Black and Latinx students—with well-qualified and well-supported teachers, adequate space, and reasonable class sizes—is not a priority for those in power. This is clearly an example of both racism and classism that is enacted through a disinvestment in public education.

Inequities in Access to Rich Instructional Resources and Technology

What does access to rich instructional resources and technology look like specific to in-class mathematics instruction? In addition to having a well-trained seasoned mathematics teacher with both the background and the experience to effectively engage students in the learning of mathematics, to teach mathematics effectively one can rely on various physical instructional tools, including manipulatives, technology, vertical non-permanent surfaces (VNPSs) such as whiteboards to make thinking and learning visible (Liljedahl, 2020), and adequate textbooks or other instructional resources such as time for planning, opportunities to work with one's peers, and access to mathematics coaches and specialists. Specific to educational materials, at the lower grade levels manipulatives include Unifix cubes, Cuisenaire rods, shapes, solids, puzzles, and many other educational aids. At the high school level these include three-dimensional models as well as algebra tiles, balances, protractors, and compasses. Imagine how much more effective it might be to study the properties of a three-dimensional object—say, a pyramid with a square base—if one has a model of that object in one's hand. You can pick it up and look at the shapes that you see. You can run your finger along the edge that connects two of the triangular faces. With the pyramid in your hand, you can touch the point at which the triangular faces meet and move the pyramid around to see the other vertices more easily than if you were looking at a picture. There is a beautiful relationship between the number of edges, vertices, and faces of certain three-dimensional solids. Specifically, this holds for convex *polyhedra*—that is, three-dimensional shapes with straight segments for edges where if you connect any two points the line segment drawn is fully inside the figure (cubes, pyramids, prisms, etc.). The number of vertices (V), edges (E), and faces (F) are connected by this formula: $V - E + F = 2$. This beautiful relationship was discovered by the Swiss mathematician Leonhard Euler in the 1700s. Imagine learning about this formula with a number of solids in your hand. You can count the number of vertices, edges, and faces and, by exploring a number of these solids, begin to get an intuitive sense of the relationship between them. This is much harder to do using flat pictures instead of these three-dimensional objects.

With respect to technology, mathematical resources might include access to learning software specific to mathematics—such as programs like *Desmos®* and *GeoGebra®*—as well as calculators, computers, and robotics equipment. A lack of resources leads to less engaging lessons. Liljedahl (2020) examined the classroom conditions that foster student thinking. He found that having students stand up and work at VNPSs dramatically increases student thinking, collaboration, and the mobility of knowledge around the room. Quite literally, this simple resource has a profound effect on how students engage with mathematics. Mathematical resources also include textbooks or other up-to-date instructional resources. Schools with more resources are better able to acquire and maintain texts and materials that are aligned to the most recent standards and with the most recent standardized exam, as well as supplemental texts, which provide for a richer experience.

The inequity that exists across schools with respect to instructional resources and technology was made extremely clear when, in the spring of 2020, K–12 schools moved to distance learning as a response to the COVID-19 pandemic. As COVID-19 case counts and death rates rose, in many states and provinces public schools were mandated to shift instruction from in-person to distance learning, at least for a time. In most states and provinces, private schools did not have to abide by the same requirements as public schools since they do not receive funding from the government, save for universal pre-K programs in some cases. Government decisions around school closures and the ability to ensure students had viable supports for learning remotely were two major factors that influenced how students fared in their learning. Let's look at a particular case in more depth.

As cases of the virus rose fast and early in New York, the state government enacted the New York State on Pause order, which, among other things, shifted instruction from in-person to distance learning. In March 2020, as both cases of the virus and the possibility that the city would be shut down grew, private schools (which include parochial and religious schools) in the state began to prepare for such a move. Families were asked whether students had access to electronic devices such as computers and iPads. Those who did not were sent home each day with a laptop in case the schools were closed to in-person learning the next day. Teachers participated in trainings around the use of distance learning. In classrooms, teachers walked students through the various websites and systems they would need to access to participate in distance learning. The schools held a practice day where instruction was conducted exclusively online but in the presence of a classroom teacher so that students could have the support needed should they run into technical issues. Passwords and usernames were shared with families, and there was a flurry of communication from the schools with respect to online instruction. When classes moved online, the transition was basically seamless.

By contrast, though it seemed to many that schools were headed to closure, the New York City mayor at the time, Bill de Blasio, insisted that they would remain open. As a result, the public schools did not adequately prepare for a move to remote learning, choosing instead to continue to teach in the manner they had prior. While teachers and administrators in many schools spoke up, fearing a lack of preparedness should the schools shift to online instruction, they were instructed to, against their better judgment, continue to teach in the manner they had been doing. Schools closed to students on Monday, March 16, 2020 (teachers came in later that week for trainings). Students were sent home with various books but without electronic devices or the usernames and passwords needed to engage fully with online instruction. In the days and weeks that followed, teachers reached out to students and their families by phone and email, attempting to walk them through setting up and accessing the websites and systems they would be required to use during remote learning. As many students did not have the technology needed, this was a difficult and, at times, futile endeavor. The city began to distribute electronic devices to students, but it was estimated that as many as 300,000 students were without the devices needed to effectively engage in distance learning in mid-April (Feiner, 2020), roughly one month after schools moved to remote learning. Unlike most private schools, where there is a laptop for every child, the public schools were looking to companies like Apple and T-Mobile to provide iPads to students, causing a delay. If these schools had had the resources they needed from the start, all students could have had a device from the very first day of remote learning, but a constant underfunding of city schools has led to a large disparity between the resources available to them and those available to schools serving their peers from more affluent areas. The move to distance learning for the K–12 public schools was an utter disaster. The difference in preparedness, resources, and planning between the public schools and the private schools led to very different transitions to remote learning. In doing so, students in the public schools suffered through a poorly coordinated effort that disrupted their education.

Of course, the case highlighted here is one of many. Even among public schools, there were differences between the smoothness of the transition and the access to resources students had. In many cases these differences highlighted disparities in resources with those schools in more affluent areas having a smoother transition to remote learning than those in poorer areas. Access to technology and other instructional resources is not plentiful at schools that are strapped financially. Students in poorer areas may not have necessarily had the technology they needed at home. Further, even with the technology many did not have adequate internet access in their homes and as such had a hard time accessing their education during this time.

Inequities in Access to Advanced Coursework

It isn't just a lack of physical and personnel resources in schools that serve predominantly Black and Latinx students; there is also a persistent lack of opportunities in terms of advanced coursework. High schools that serve predominantly Black and Latinx students tend to have fewer AP courses, and these schools, at all levels (K–16), offer more remedial courses. As an example, according to data compiled by the U.S. Department of Education Office for Civil Rights (2018), while 50% of high schools offer calculus courses, the percentage drops to 38% when considering schools with high Black and Latinx enrollment. These schools are less likely to have a student mathematics team, to participate in mathematics competitions, and to offer enrichment programs in science, technology, engineering, and mathematics (STEM). Labs might be less modern, and the existence of STEM labs less prevalent. These schools tend to offer more remedial classes and fewer enrichment opportunities. These schools also suffer from high teacher turnover, overcrowding, and counselors with much higher caseloads. It is estimated that 60% of fourth graders and 66% of eighth graders across the United States during the 2019 NAEP exams attended public schools with overcrowded classrooms (Cai, 2021). Often these are schools that serve students in the poorest of areas.

Inequities in Community Supports

The COVID-19 pandemic highlighted the inequities in communities and home lives that exist between students from poorer backgrounds and their peers whose families are more affluent—inequities that have a direct influence on math learning, including access to nutritious food, a quiet and safe space to think and work, health care and mental health support, and access to special education and other therapeutic resources. First, as schools closed, access to free and reduced-price breakfast and lunch programs dwindled. This meant that many were without the meals they had come to rely upon. In fact, approximately 30 million children in the United States receive a free or subsidized lunch through the National School Lunch Program each year (Purvis, 2021). Students whose household is living below 130% of the poverty line are eligible. In an effort to provide meals, school sites designated times when families could pick up meals, but as people were "social distancing" and others did not have the time to pick up meals because of having to work all day, these methods were not very effective at getting food to those who needed it, and the result was students suffering from food insecurity. In a School Nutrition Association survey of 1,894 school districts, it was found that 3.3% of school districts were not serving meals and an additional 1.7% were offering meals at one point but stopped. Of the districts that were serving meals, over 80% stated that the number was far less than the usual number they serve, with 23% of districts stating they served between 75% and 100% fewer meals than before the pandemic (School Nutrition Association, 2020).

Second, when remote learning began, in many places it became painfully clear that not all students have a quiet space in which to work. Having multiple children at home trying to attend school remotely highlighted the fact that some live in more crowded spaces that affect how they learn. There were stories of students connecting to their classes in bathrooms and closets as those were the quietest spaces available in their homes. Some students had to take care of their siblings while their parents went to work, further hindering their ability to engage in classwork. This was more prevalent among families whose members held jobs that did not allow for remote working situations such as delivery workers, transit workers, and those in health care fields. Additionally, students who receive special education services missed out on critical resources including working with special education teachers in structured environments. With special education services moved online, providers did not have the ability to use the tools and resources they would normally use with students in spaces specifically designed for their use. As a result, students received a less-than-optimal experience.

Access to health care and mental health support for students and their families was another area of vast inequity. A review of 10 studies on the impact of COVID-19 on student health and well-being by Chaabane and colleagues (2021) indicates that, as expected, school closure also led to a rise in anxiety, loneliness, stress, and sadness among school-age children. Many children were traumatized by illness and death in their families and in their communities, many of whose access to medical care was not as strong as it should be. And lack of affordable access to school counselors and other mental health professionals left many children and their families on their own to fend for themselves. As schools returned to in-person learning and students were being encouraged to accelerate and catch up to grade level, what messages were they being given to help them feel confident and competent in their mathematics learning? How much *more* "bad at math" did many of these students feel? There is a myriad of challenges that affect daily the educational lives of our students (access to food, health care, and mental health supports, a safe and quiet learning space, responsibilities to care for siblings, etc.). These opportunity gaps drive the outcomes we see and only serve to continue the pervasive "bad at math" sentiment.

Inequities in Budgets

It is estimated that 47% of K–12 education funding in the United States comes from state revenue, with 45% being provided by local government and 8% by the federal government (Partelow et al., 2018). Between 2008 and 2018, in part because of the recession, state funding to schools declined by 20% (Partelow et al., 2018). This led to schools having to cut their budgets, which resulted in fewer resources, fewer enrichment programs, and less support for the education of our students, most especially for schools that serve students from lower-income backgrounds. Funding matters when it comes to education. The decade following the great recession was a time when per-pupil spending suffered its largest and most sustained decline in over a century. A study found that states with larger budget cuts experienced a decline in scores on the NAEP exams during this time (Jackson et al., 2021). Specifically, a 10% spending cut was associated with a drop in NAEP test scores by 7.8% of a standard deviation and a 2.6-percentage-point drop in graduation rates. To ensure that the results did not simply reflect the effects of the recession in general, the study compared states that were more reliant on state revenues for educational funding with those that were less reliant on these resources and as such experienced different degrees of cuts to educational funding during this time. It found that test scores declined most in states where funding was most drastically cut. Additionally, as the economy improved, the cuts remained, and the lower scores did not recover. This drop in scores marked the first time in over 50 years that there had been a drop in scores in both mathematics and English on the NAEP exams, and scholars believe it was driven by the cuts to the education budget. Further, the rate of first-time college goers also declined 2.8 percentage points for every $1,000 less spent on per-pupil funding during the time that scores on the NAEP exam dropped.

The recession coincided with a drop in the share of state revenue going to fund public colleges in all 50 states and an increase in tuition at public colleges in 47 states (Baylor, 2014). What this means is that at the college level, a disinvestment in public education over the past decade has led to more of the burden of the cost of obtaining a college education landing on the students themselves, leading to an explosion in student debt over this time. Public colleges facing budget cuts have increasingly relied on tuition revenue, which is placing a great burden on families with lower incomes. Presently, there is over $1 trillion in outstanding student loan debt in the United States (Baylor, 2014). The share of students in public colleges borrowing money for tuition increased by 6% between 2008 and 2012 alone. The total amount borrowed at public colleges through federal student loans increased by $17.1 billion when comparing the 2007–2008 academic year to the 2011–2012 academic year. This is a 54.6% increase. Student borrowing increased most, as expected, in states that made the biggest cuts to per-pupil spending. The disinvestment in public education has contributed greatly to the burden placed on Black and Latinx students with respect to student debt. It has, further, deprived those same groups from the resources, facilities, and opportunities needed for those in the K–16 school system to thrive.

WHAT CAN YOU DO TO CHALLENGE THE OPPORTUNITY GAPS THAT EXIST?

In this chapter, we saw how stark differences in opportunities lead to differences in outcomes. Here the mandate is clear: We need to reverse the chronic underfunding of public schools. We need to attend to buildings and facilities that have been neglected, increase access to supplies and resources, and ensure that our schools are fully and adequately staffed. We need to build more schools to eliminate overcrowding in our major urban centers. The conditions in those schools we consider our nation's best should be replicated in all schools so that students have access to the conditions necessary for their success. This requires a commitment among those who work in government and who control budgets to invest in our schools. It also requires a commitment among voting citizens to support policymakers who are committed to this cause.

It may seem that these goals are out of reach and perhaps beyond what a classroom teacher, instructional leader, or administrator can control, and yet, here, too, you have a role to play. You can help propel these issues into the spotlight not only at your school but also at the regional and national levels. Consider raising awareness about the inequities you see by

- writing a letter to the editor of your local newspaper about inequities that you see and their impact on your students, and sharing any ideas you have for addressing these;

- regularly attending and speaking out at school board meetings with respect to the needs of the students you work with;

- joining an education advocacy group to bring awareness of issues to a greater audience;

- calling out opportunity gaps when you see them, including in your conversations with colleagues and parents;

- pushing back against achievement gap rhetoric when you hear it; and

- committing to raise awareness of inequities—not just when the situation becomes so dire that it can no longer be withstood, but always.

Questions for Reflection

For Teachers

- How can you raise awareness about inequitable situations in your school?

- What instructional resources are available to you at your school, and how can you leverage these to improve the teaching and learning of mathematics?

- In what ways can you advocate for additional materials such as graphing calculators, Desmos®, VNPSs, and other resources to enhance your teaching if they do not already exist at your school?

- How might you advocate for additional support in terms of human capital (math coaches, instructional leaders, etc.) or step into such a role yourself?

- How can you engage with professional organizations and educational activist organizations to push for a more equitable system of public schooling?

- In what ways are you sensitive to students' experiences and life situations as you work with them?

For Instructional Leaders

- In what ways can you support teachers in raising awareness about inequitable situations in their schools and the greater system of schooling?

- How might you ensure that the teachers you work with make full use of the instructional materials and other resources, including time and access to key personnel, that they have at their disposal to enhance the teaching and learning of mathematics?

- How might you support them in acquiring additional materials so they have what they need to equitably serve students?

- How can you support teachers who wish to step into leadership roles, lead professional development, or otherwise engage with their profession in new ways?

- How can you engage with professional organizations and educational activist organizations to push for a more equitable system of public schooling?

For Administrators

- How might you use your power and privilege to raise awareness about inequitable conditions in the school system?

- In what ways can you leverage grant writing to increase resources at your school?

- How can you ensure that existing budgets are equitably distributed at your school?

- How might hiring additional personnel, including math specialists, afford teachers more time and support for planning, teaching, and working with others?

CHAPTER 9

IS THE SCHOOL SYSTEM BROKEN?

In this chapter we will:

- Consider and challenge the argument that the public school system is broken.

- Explore the differences in how schools serve students based on the historical development and purposes of public education.

- Consider multiple purposes for schooling.

- Explore teacher professionalism.

- Reflect on ways to expand teachers' roles and influence in the educational system.

IS THE SYSTEM OF PUBLIC SCHOOLING BROKEN?

Whether the present system of public education is broken is a matter of what one chooses to measure. In many ways, the school system is doing an excellent job of the function for which it was primarily designed. The system of public schools is a little over 200 years old and came into existence out of a need for workers. Industrialists needed workers for their factories. As such, they had a large influence on the design and curriculum that was established. These individuals were looking for workers who were punctual, disciplined, and able to focus on the task at hand for long periods of time, as well as able to follow instructions. In a factory environment, the abilities to pose thoughtful questions about the tasks, to critique, and to think critically about one's role and the system in which one is employed are not, from the industrialists' point of view, good characteristics for their workers to develop. As a result, school systems were developed where rote memorization and the following of rules and procedures were paramount. Critical thinking and the ability to question were downplayed, as the purpose of the schools was to create good, docile, obedient workers. This was as evident in mathematics education as in every other subject. The strength of the U.S. economy and the expansion of capitalism are evidence that the schools are indeed succeeding in providing a solid group of dependable workers year after year.

WHAT WAS SCHOOL DESIGNED FOR?

To understand why we see such disparities within our school system, we must look at some of the historical factors that led us here. For example, at the dawn of the public education movement in the United States, the children of the wealthy factory owners were usually attending different schools than the workers they employed. This led to differences in curricula and in focus across schools that served students from different social backgrounds. This difference persists today, and was made exceptionally clear by educational researcher Jean Anyon (1997) in *Ghetto Schooling: A Political Economy of Urban Educational Reform*. In the text, Anyon explores differences in how schools that serve students of differing social statuses prepare their students. Studying schools in Newark, New Jersey, she was able to show how schools that serve students from working-class backgrounds value obedience, rule following, and the memorization of facts over critique, questioning, and the understanding of broader social systems as schools serving those from higher socioeconomic backgrounds do. As a result, the education received by some students is quite different from that received by others. Some students are being prepared to work, while others are being prepared to lead.

> **Some students are being prepared to work,
> while others are being prepared to lead.**

Such a differential system serves also to continue the present society itself. Nations teach their young, in part, so that they may continue to value and work to maintain the nation itself.

The fact that large systems—including the educational system—reproduce themselves has been written about extensively. The philosopher and researcher Pierre Bourdieu and his colleague Jean-Claude Passeron (1990) write about how the system of education serves to reproduce the inequities already present in society. That some schools are so much better funded and resourced than others and provide a distinctly different set of experiences for students means that schools serve to sort students, putting before them possible paths for their future. We have already seen how an overreliance on mathematical testing affects access to future careers and schooling. Mathematics education, unfortunately, has traditionally aided in the process of sorting.

Jean Anyon's (1980) text, *Social Class and the Hidden Curriculum of Work*, clarifies this feature of schools. In it, she describes five schools. She calls the first two working-class schools. Here, most of the parents have blue-collar jobs. Men in the households tend to be unskilled or semiskilled workers, and approximately 15% of them are unemployed. Fewer than 30% of women work, some of whom are working part-time, and approximately 15% of families live below the federal poverty line. The third school is described as a middle-class school with a mix of workers represented among the parents: skilled blue-collar workers who are well paid, as well as working-class and middle-class white-collar workers. The fourth school she labels the affluent professional school, with most of the families having parents who work in professional industries. The school is also 90% white. The fifth school she calls the executive elite school. There, most of the men in the households are top executives in major U.S.-based multinational corporations, with many working in top financial firms. Some of the women in the households are also employed in well-paying elite jobs, and many volunteer in service groups and participate in local politics. Family income is quite high and was representative of only 1% of the U.S. population at the time the article was written.

Anyon (1980) compared the fifth-grade classes of the five schools, and while she found similarities in the instructional materials used, she found differences in how the curriculum was enacted. In the two working-class schools, rote memorization of facts and the following of procedures was highlighted. Students were told the steps, teachers wrote the steps on the board, and students copied them. Students work was graded less based on whether it was accurate and more on whether they followed the steps. They were not encouraged across a range of subjects to be creative, draw upon data or texts to come up with their own conclusions, or share their thoughts. Classroom space was controlled, and decisions about the space and how time would be used were made solely by teachers and administrators. Also, the texts that were available were often not used, as teachers preferred to use worksheets. In the middle-class school, schoolwork involved getting the right answer, and while following directions was valued, there was usually some figuring, choice, and decision making that went into the process. The text was featured in most lessons but was not analyzed critically; rather, it was taken as fact. In the professional school, work was a creative activity, often carried out independently. Students were asked to express and apply ideas and concepts. The products of this work were often written stories, editorials, or artistic representations. There were fewer rules and more acceptance of individuality in one's work. Work at the executive elite school consisted of developing one's analytical and intellectual capabilities. Students conceptualized rules by which elements fit together within systems and then used these rules to reason through problems. Students were asked whether they agreed with the reasoning presented by their peers and the texts alike. Further, movement was not as restricted, and students moved around the room freely to get their materials. After a few months of school, they moved from class to class freely without lining up and often without their teachers' presence, unlike in other schools.

What Anyon (1980) concluded from this is that students at the different schools were being prepared for different kinds of work and to have different relationships to capital. The students at the working-class school were learning to follow directions, to not question authority, and to complete repetitive, rote tasks in a compliant manner. In the middle-class school, students were learning skills for white-collar jobs where knowing the right answer is not as necessary as knowing how to find the right answer. This is preparation for technical work, paperwork, and other tasks in a bureaucratic system. In the affluent professional school, students were developing the ability to be expressive, to negotiate, and to develop their ideas in their own ways. These are skills needed to produce culture. In the executive elite school, students were being taught to think in terms of systems and to be able to analyze systems in a way that none of the other schools did at all. All of this produces workers with very different skills, relationships to capital, and future paths as far as careers are concerned. These reinforce and maintain the current social system where each of these students has a place defined in great part by the school they attend.

When I first read Anyon's work, I often thought about my own education and the types of future work the various schools I attended were preparing me for. I found I could relate to her work and saw parallels in my own schooling. Despite the fact that the research is, at this point, more than 40 years old, I still find it a compelling argument. Do you feel this research still holds true today? What, if anything, has changed? How, if at all, do mathematics classes now reflect these differences in preparation?

QUESTIONING THE EXISTENCE OF A MERITOCRATIC SYSTEM

If we consider the role of schools as both producers of workers and reproducers of society, tasks for which they were designed, then schools are not indeed failing. Further, in nations such as the United States that are built on systemic racism, the reproduction of inequality has involved the teaching of stories that maintain the superior positioning of whites as compared to other racial groups. Many of the stories of Black and Latinx individuals have been discarded, hidden, and reimagined to maintain a view of our society as meritocratic, where individuals can achieve their dreams if they persist, persevere, and work hard. We buy into the ideal of meritocracy when we think of Black students as struggling more in mathematics than their white counterparts due to a lack of effort or discipline rather than as a result of differences in the opportunities, supports, and resources available to them. We buy into the ideal of meritocracy when we attribute a Latinx family's lack of attendance at a parent–teacher night to their not valuing their child's education, as opposed to any number of other quite possible explanations—many of which again relate to opportunity, support, resources, and in some cases cultural beliefs and understandings. The belief in a meritocratic system of education and opportunity in general is supported through the telling of stories of individuals who despite the difficult circumstances in which they grew up were able to succeed at the highest levels. Such stories keep alive the idea of a meritocracy, as some conclude that if the individuals in such stories were able to succeed, then the possibility exists for all to do so. The trouble is that these cases are rare. They are not representative of the opportunities available to all, and our belief in them narrows the conversation around access and opportunity. It keeps us from questioning why students are getting access to such differential school experiences and what that access is preparing them for. It keeps us from questioning why some students have access to high-quality rigorous instruction in mathematics while others do not. That a handful of individuals transcended a lack of opportunities to become wildly successful should not validate a system that fails to provide opportunities to all individuals in an equitable fashion.

> That a handful of individuals transcended a lack of opportunities to become wildly successful should not validate a system that fails to provide opportunities to all individuals in an equitable fashion.

It should not keep us from questioning both the educational system and the wider opportunity structures within our social world. These cases, though exciting and worthy of sharing, are often used to justify the continuation of processes and structures that are, in and of themselves, inequitable, racist, and sexist.

SCHOOLS AS CHALLENGERS OF INEQUALITY

I write this not to downplay the benefits of education to the lives of individuals, especially those from marginalized communities. Educational attainment is certainly linked to an increase in traditional measures of success. School matters in the lives of students, and yet we cannot ignore the differential access to high-quality education that exists in our society. Imagine how much better the lives of our most vulnerable students might be if they had access to high-quality schooling and experiences that led to questioning, critique, and the development of critical thinking skills. Imagine how much more exceptional society itself would be if those same students were able to think about large systems, their role in the society itself, and questions of equity, fairness, and access. Might we be able to challenge the inequities that exist and dismantle the sexism, racism, and numerous other isms that corrupt the lives of so many?

We have seen in this chapter two purposes of schooling that in many ways serve to perpetuate inequality. We can consider and reimagine schooling in ways that highlight other goals, serving instead to challenge inequality. One of these goals centers on developing a student's full potential. If we look at each student as a unique individual full of talents, gifts, and abilities, then one possible role of education is to develop these talents and abilities to their fullest. In this way, education becomes about personal fulfillment and ensuring that all students are supported as they grow and develop. Supporting each child as a unique individual includes ensuring that they have access to the most supportive educational experiences possible and necessitates our addressing issues of unequal access and differential educational experiences.

A second goal for schools that serves to challenge inequities focuses not on the individual but on the society. This additional goal is to improve the society itself by developing in students the potential to understand complex social systems as well as to understand and challenge systemic racism and all barriers to the creation of a more equitable, more just social world. This does not mean that we all must be on the same side of every issue or that we must all agree on the best way to achieve equity, but it does mean that the commitment to equity must be present in the system of schooling. Then as students become full, participating, and engaged members of society, they are fully prepared to tackle issues of inequality however they choose to do so. It further means that schooling must include an examination of one's place in the social system. A raising of consciousness about the ways in which schooling itself has been used to maintain inequities must form part of our system of schooling if we are to truly challenge the injustices still present in society. The Brazilian educator and social activist Paulo Freire (1994) referred to this process as *conscientização*, or *critical consciousness* in English. It is a focus on understanding the social and political realities of the world in all their complexities so as to improve one's own position within it. Those oppressed by society can then use their learned social and political knowledge to challenge that oppression and to liberate themselves from it. Schools can thus serve as a place both for individual and for social transformation. Mathematics can serve as a tool for understanding one's social position in the world and to advocate for social change toward a society where there is more opportunity for all. One way to do this is through the teaching of mathematics for social justice, which we will consider in Chapter 10.

TEACHER PROFESSIONALISM

One ally in such work must be teachers, as they are both experts in education and most uniquely positioned to know and understand their students. They spend many hours with students and are intimately involved in their studies. Yet classroom teachers in the K–12 public school system, for all their training and preparation, are kept from fully utilizing their professional discretion when it comes to educating students. The biggest difference I have found between teaching at the high school level and teaching at the college level is the amount of autonomy one has and the degree to which one is treated as a professional with respect to pedagogical choices. At the K–12 level, it is almost universally true that teachers are given the curriculum they are to use. They do not pick the textbooks, do not determine the pacing, do not always design their own assessments, and are often told, in excruciating detail, how to organize their classroom instruction. By that I mean that they might be required to use a mini-lesson format, to include three to five questions of a certain type, to use a strategy such as think-pair-share, and to incorporate certain specific pedagogical and technological tools that they do not have a hand in selecting. In some cases, teachers are given a script to follow when teaching. They are literally told what to say. Elementary school teachers are even told how often to change their bulletin boards, what color backgrounds to use, and how to organize them.

At the college level, by contrast, one is given the name of the course and the course description from the bulletin and, except for some multi-section 100-level courses, told to run with it. College professors choose the texts and readings that will be used in their classes. They are in complete control of how their lessons are constructed, how the class will unfold, the assessments and activities to be used, the technological or other tools to be incorporated, and the pacing to be followed. As long as the course adheres to the description of the content, pacing and delivery of the content are left to the professor.

What can be concluded from this is that teachers, unlike college professors, are not treated as professionals when it comes to teaching their courses. Professionals have the autonomy to make decisions about how to best carry out the responsibilities of their jobs. They use their extensive training and relevant research to make decisions that impact the work they do in tangible ways. Lawyers, as an example, are not given a written script to recite in court or a predetermined argument to follow with respect to a particular case. They take the information in front of them and their knowledge of the law to construct the best argument they can using their knowledge, skills, and experience. Similarly, doctors diagnose patients and recommend treatment plans while relying on their training, experience, and research. Teachers, on the other hand, while trained as educational professionals, are not given the same professional autonomy.

THEORIES OF PROFESSIONALISM

Work on theories and models of professionalism has identified several pillars of professionalism that include training, initial practice, and credentialing; ongoing development to keep abreast of changes in the field; having members' judgments be well regarded both within and outside of the field so that everyone has a voice in how the field should grow and adapt; and active discussion and collaboration among members (Abbott, 2014; Ingersoll & Collins, 2018; Organisation for Economic Co-operation and Development, 2019). Further, Mirko Noordegraaf, a professor at Utrecht University in the Netherlands, notes that professions require professional control. That is, "professionalism demands the capacity to internally organize and protect 'professional practices from external influences'" (Bruno, 2018, p. 2). Some of these certainly do apply to teachers, but not all do. From this we can discern that teaching is a semi-profession with teachers certainly being required to receive licenses and having initial practice and training required of them, but without their voices and judgments being well regarded, without a voice in how the field should grow and adapt, and without the ability to control their profession. According to Hargreaves (2000), teaching has passed through four historical stages: pre-professional, autonomous, collegial, and the stage we are currently in, which is termed post-professional. In this last stage, while teachers exhibit a commitment to service, they are very much externally controlled through prescriptive policies, outside evaluation systems, and external control of the profession by those other than teachers themselves. Over the past three decades, a business model of education has developed where the focus has been on outcomes without consideration given to the inputs. Students, teachers, and schools are evaluated based on how well students perform on externally designed standardized tests without regard to local conditions or the opportunity structures within and outside of schools, and without input from teachers on the standards to be followed, the curriculum to be used, and the methods and instruments of assessment. We have moved to what Zeichner (2019) called a managerial or organizational model of professionalism where market and business models have been brought to the educational sphere. These have emphasized privatization, corporate-run schools, mayoral and state control of school districts, vouchers, high-stakes testing, and the reliance on standardized tests for evaluation. With these changes has also come a more prescriptive and rigorous set of requirements for teacher licensing.

TEACHER PREPARATION

In the United States and Canada, teachers are required to be licensed to teach in the public schools. The requirements for licensure vary across geographic regions (states or provinces) but tend to be rigorous. They include specific coursework in education as well as content courses, fieldwork hours, teacher licensing examinations, and practicums. Further, after teachers obtain their license, they are required to complete ongoing professional development to maintain their license and remain current in the field.

Over 91.9% of teachers participate in ongoing professional development. Yet when it comes to the types of ongoing experiences that would be most beneficial as highlighted in the research literature (visits to other schools, university coursework, and the ability to present at conferences), only about one quarter of practicing teachers are supported in carrying out such experiences (García & Weiss, 2020). Reimbursement for college courses is rare, as is being given reassigned time from teaching to participate in site visits, speak at or attend conferences, and take part in other professional opportunities. Further, building time and opportunity into teachers' schedules for them to collaborate, discuss the profession, and support one another rarely occurs. Teaching is still very much an isolated practice.

The expectations for teachers in terms of preparation, licensing, and ongoing professional development are high. Education departments and boards lay out expectations for what is considered a profession, and yet when it comes to the classroom, there are many decisions about the education of their students that are left out of the hands of teachers. This is not because their preparation does not warrant their being afforded professional autonomy. If education departments and boards were unconvinced that teachers were well prepared to make such decisions, they could alter their already rigorous licensing requirements as has happened multiple times in the past. Yet, even after meeting these requirements, teachers are not given the professional autonomy necessary to develop and implement curriculum and materials of their own in teaching.

One area that bears discussion with respect to teacher preparation is the teaching of mathematics at the elementary school level. Teachers at this level tend to be generalists who do not necessarily have a thorough preparation in mathematics. One way to strengthen the teaching of mathematics is to have mathematics specialists who have been prepared to teach elementary mathematics specifically teach mathematics to students at this level. The National Council of Teachers of Mathematics (NCTM) supports a reliance on mathematics specialists at the preK–6 level, suggesting these individuals could teach mathematics at this level and also work with other teachers to strengthen their ability to teach mathematics. This is evident in a joint statement NCTM (n.d.) put out with the Association of Mathematics Teacher Educators (AMTE), the Association of State Supervisors of Mathematics (ASSM), and NCSM, Leadership in Mathematics Education. We are seeing teachers of mathematics specifically identified for students as young as fourth grade in some school districts, and we have seen the development of teacher preparation programs at the college level to train such elementary math specialists. This is a welcomed change but one that is happening too slowly.

TEACHERS AS SCAPEGOATS

What troubles me most with respect to the teaching profession is the reality that while they are not in control of many pedagogical choices and are constantly being forced into particular ways of teaching, teachers are consistently held responsible for failures big and small in the school system. It seems quite a fallible argument to hold K–12 teachers accountable to such a high degree while simultaneously limiting their autonomy, but that is exactly what happens. Why, then, is this argument so prevalent? Simply put, blaming teachers narrows the conversation, shifting our focus away from the larger issues facing the school system and society at large. Teachers become scapegoats.

The issues affecting student performance go well beyond the classroom walls. Whether a student's family has stable appropriate housing, access to medical care as needed, food, and other basic needs met impacts a student's ability to learn. These are factors well outside a teacher's control. Many of these, including financial stability, access to housing, and adequate medical coverage, can be strengthened through legislation and social programs. The issues present in housing, health care, and employment are outside education and yet impact what happens in schools in very real ways. Ensuring that there are adequate jobs, care, and affordable housing for families in our society would strengthen the schools as well. It is, of course, much easier to avoid addressing these, given not only the large amount of work needed to reimagine these systems so that they more equitably serve students of all socioeconomic backgrounds but also the sustained commitment to equity and financial investment required for such a reimagining to be put into practice. Trying to change the school system without dismantling the systems of oppression that led to its inequitable conditions in the first place will yield little to no results.

> Trying to change the school system without dismantling the systems of oppression that led to its inequitable conditions in the first place will yield little to no results.

Until we recognize that the teacher is working within a school, a school system, and a society whose institutions support the inequities we see reflected in schools, we cannot truly expect to make lasting changes that benefit our most vulnerable students.

The necessity of such work becomes less clear when we instead center failure on teachers. Doing so forces upon us different ways of responding. We bring in new curricula, new programs, new technologies, and seemingly endless amounts of professional development. We then seem surprised when these do not work as expected, and we restart the process. If we consider who benefits in all this, it is the companies that make these products as they profit from the constant cycle of adoption of new materials. Teachers who become comfortable with one set of materials are made to move quickly to another, and unfortunately their students do not reap the benefits of having a teacher who is well versed in the materials being used.

The blame-the-teacher mentality is wrong for numerous reasons. First off, it is flat-out inaccurate to assume that the trouble students have with mathematics or with academic attainment in general can be explained solely by their teachers. In a complex educational system within a larger society rife with inequality, sexism, and racism, the idea that responsibility for the persistent underservicing of Black and Latinx students rests solely on the teachers is simplistic and erroneous. More than that, however, this thinking pushes us solely in the direction of reforms that center on classroom instruction. It points us to the narrowest of solutions. It keeps us tinkering within the same system, trying to refine and revise it by focusing on ways we can better develop teachers and when we should be stepping back, zooming out from the classroom level, and looking at *all* the changes necessary to improve our society and, in doing so, our schools.

All this is not to say that the professional development of teachers is unnecessary or yields little results. Teachers matter. A qualified, dedicated, and well-supported teacher can make all the difference in a child's life. Providing effective teacher training is part of the work of supporting teachers. However, by focusing on this exclusively, we lose sight of the bigger picture. We find ourselves in an endless loop of changes. Many of these are both unnecessary and taxing on teachers who, in a short period of time, are being asked, yet again, to try something new but not being invited into the discussions and decisions around what curricula and materials to use and what pedagogical approaches make most sense to them and their students. At the classroom level, the teacher is the professional in the room. We should start to treat them as such.

> At the classroom level, the teacher is the professional in the room. We should start to treat them as such.

TREATING TEACHERS AS PROFESSIONALS

What does it mean to treat teachers as professionals and to treat teaching as a profession? To begin with, we could step back from the constant micromanaging and give teachers the autonomy and responsibility to make decisions that affect classroom instruction. This involves decisions about pedagogy as well as decisions about curriculum, pacing, and resources. It should also include the opportunity to design assessments and implement a robust, holistic approach to assessing students that gives us more information about their abilities and growth than one-shot high-stakes exams do.

Next, teachers need authentic opportunities for reflection, collaboration, and development. Time should be built into teachers' schedules for meeting with other teachers to design lessons, curriculum units, and assessments, as well as to share ideas and support one another. Observing other teachers can be a powerful way to reflect on one's teaching and learn about one's profession, yet this is rarely done in schools. Building in time to do so is of value. Additionally, teachers should be part of the decision-making team at their schools, and opportunities should be created for them to take active roles in designing and carrying out the mission and vision of their school including decisions around governance. To involve teachers in decision making around all aspects of the work of the school is beneficial because the work of schools is carried out primarily by teachers, and their voices and experience can aid in ensuring a successful environment for students to learn in. Top-down mandates that are imposed on teachers can be challenging to work under and undermine the work of schools. Involving teachers in decision making, running ideas and policies by them, and having their ideas be part of the conversation help to counter this reality, though that is not often what we see in the school system at present.

Professionals also require the ability to improve their craft and grow in their profession. In addition to opportunities to observe each other and work together within schools, time should be built into the school year for teachers to attend professional development that goes beyond one- or two-day workshops. This might include both the time and funding to attend educational conferences and to take courses at the university level. Opportunities can be developed that afford teachers the ability to apply for a paid sabbatical on a regular schedule that includes funding for coursework and ongoing professional development. Memberships to professional organizations should be provided to teachers as these often come with subscriptions to academic and pedagogical journals that they would benefit from reading, access to online resources, and opportunities to participate in workshops, webinars, and conferences. In mathematics, the NCTM—the largest professional organization in the United States and Canada dedicated exclusively to mathematics education—offers numerous benefits including a practitioner journal, an academic journal, and access to local, regional, and national conferences as well as an incredibly useful website. Local and state-wide professional organizations also exist, which teachers would benefit from being a part of. Perhaps teachers could choose which organizations they wish to be a part of, and some or all of their dues could be covered by their schools as part of efforts at developing teachers professionally.

To share your expertise in the classroom:

- Consider collecting student work and writing about the lesson you taught that led to it. Submit your writing to a teacher journal such as the NCTM publication *Mathematics Teacher: Learning & Teaching PK–12*. Instructions on how to submit your writing there can be found at www.nctm.org

- Consider collecting student work and sharing it along with the lessons and activities that led to it at local, regional, and national conferences.

While some teachers do share their lessons, activities, and experiences with others by leading professional development sessions and giving talks at conferences, this is not a widespread practice. For teaching to be viewed as a profession, teachers should be encouraged to share their experiences broadly to influence the profession. This includes writing articles for practitioner journals, leading professional development, and giving talks at conferences. Of course, each of these activities takes time, and to throw it onto the work of teachers above what they already do is to overwhelm them. Instead, these activities can be incorporated into regular sabbaticals. This extended time away from teaching can be used for professional development, as already discussed, but it can also be used to influence the profession through the creation of classroom materials, conference presentations, articles, and policy documents. In this way, the voices of teachers would be ever present in the profession itself, influencing the development of resources, the practices of other teachers, and the policies that guide their work. Further, teachers should be involved in decisions at the school, district, and federal levels. Decisions that affect teachers and their students should be made by or with teachers.

> **Decisions that affect teachers and their students
> should be made by or with teachers.**

By saddling teachers with the responsibility for the failures of the school system, we, in fact, preserve the successful job that this school system is doing with respect to re-creating the current unjust social system. Instead, we could be looking toward teachers as professionals with the expertise and ability to push schools forward as places where students can learn the skills needed to achieve their full potential and as drivers of social transformation toward a more equitable social world. To do so, we need to build in opportunities in terms of both time and funding so that teachers may engage in ongoing study and professional development, writing articles, developing resources, and presenting at conferences so that their work influences that of their fellow teachers and moves teaching forward as a profession.

Part of this work involves creating conditions for teachers that are supportive and provide avenues for professional growth and the ability to influence the profession. It has been argued that we do not have a teacher shortage in this country but rather that many who enter the profession choose to leave, creating what Richard Ingersoll famously termed the revolving door of teaching back in 2003. That is, while there are sufficient individuals who enter teaching, we do a poor job of retaining them. The revolving door persists to this day (Kraft, n.d.). It affects those schools that serve low-income communities most, leaving them without access to seasoned, experienced teachers. It is a problem that stems from working in settings where you do not have the resources you need, are not given the professional autonomy required, and are not treated with the respect deserved to do the job you love. This is exacerbated by a system that places blame for failure squarely on teachers, thus moving the conversation away from the real reasons why outcomes suffer. We who are aware of the misdirection of focus when this argument is used need to call its use out when we see it. Ask these questions of yourselves and others: *What do those who make the argument gain by placing blame on teachers? What does education lose when they do so?* Let us push these questions and these discussions into the national spotlight, bringing them up in settings big and small so that we can move the conversation toward what is really occurring instead of obscuring that by blaming teachers.

Questions for Reflection

For Teachers

- How can you influence the profession by sharing your experience at conferences as a speaker, by writing articles for teacher journals, or by leading professional development?

- In what ways can you take advantage of sabbaticals, if available, to further your voice, develop education materials, attain additional education, and so on?

- How might you advocate at your school and beyond for teacher professionalism?

- Might you suggest curriculum materials or develop your own for use at your school?

For Instructional Leaders

- In what ways can you support teachers in advocating for teacher professionalism?

- How can you support teachers in sharing their experiences broadly with the educational community?

For Administrators

- What opportunities exist for teachers to be part of the decision-making bodies of your school?

- In what ways can you support teachers' professional autonomy in the classroom?

- How might you use existing budgets to support teachers' travel to conferences and participation in professional organizations?

- In what ways can you leverage grant writing to increase opportunities for teachers to have the time and resources to share their experiences broadly?

- What opportunities exist for teachers to work together at your school?

- What opportunities exist for adding additional math coaches and math specialists to the staff at your school? Are there teachers already at your school who have the background and experience to take on these roles?

CHAPTER 10

TEACHING MATHEMATICS AS A POLITICAL ACT

In this chapter we will:

- Discuss the ways mathematics education is a political act.

- Consider some non-European contributions to mathematics.

- Explore the sociopolitical turn in mathematics education.

- Explore a pedagogy called teaching mathematics for social justice.

- Reflect on how you can teach mathematics in socially just ways.

MATHEMATICS EDUCATION AS NON-NEUTRAL

Many believe that mathematics is neutral and that the teaching of mathematics is, accordingly, an apolitical act. Yet what mathematics content gets taught, who is credited with the development of those ideas, whose contributions are erased, and whose get amplified are all political decisions. So too are questions of access, opportunity, and diversity in mathematics education. Further, whether the teaching of mathematics supports the current unjust social system or whether it disrupts current narratives and creates a culture of critical resistance is worthy of reflection.

One important political consideration with respect to mathematics education involves questions of access, opportunity, and diversity. We have already seen in Chapter 3 how the numbers of Black and Latinx students in gifted and talented programs in one large urban district increased dramatically when universal screening was used to determine who should be accepted into these programs (Card & Giuliano, 2016). When the exam used as the basis for admissions was given to all students, as opposed to a subset of students, the makeup of those programs with respect to race started to look more like the makeup of the district itself. Despite this, universal screening was discontinued for what was described as economic reasons. Economic reasons are often cited for political decisions such as this one, but to discontinue the program is to make a political decision about who should and should not have access to a rigorous academic program of study.

Likewise, a program was intentionally created at the University of Maryland to boost the number of students from underrepresented groups who successfully complete a PhD in mathematics. The program, which built opportunities to support such students, resulted in the college awarding three Black women PhDs in mathematics in the same year, something that had never occurred at a U.S. university before. The women—Tasha Inniss, Sherry Scott, and Kimberly Weems—received their PhDs in December 2000 (Williams, 2000). It would be beneficial, given the underrepresentation of women, Black, and Latinx students, if programs such as this were the standard at colleges across the country. It is important, also, that we change the way we conceptualize mathematics and mathematics education so that our changed mindset can drive changes to the environment itself, making such programs unnecessary in time.

NON-EUROPEAN CONTRIBUTIONS TO MATHEMATICS

Matters of policy are not the only political consideration when it comes to mathematics and mathematics education. Whose mathematics gets taught and how the contributions of diverse peoples are or are not incorporated into the curriculum are, in the end, political considerations as well. Every single civilization of which we know has a numeration system. In school you may have learned about other numeration systems. Most likely you learned about the Roman numerals and the Roman numeration system. Have you considered why, if all known civilizations had a numeration system, you learned about only one different from our own? Is it because the Roman numeration system was that much more efficient than the others? Then again, have you tried multiplying in Roman numerals? Most likely it is that the Roman empire took over large areas of land and imposed their numeration system on those they conquered. Centuries later, our society maintains a view of mathematics as the epitome of rational thought in the Western world. Part of this story is valuing the contributions of the Greeks and Romans to mathematics while excluding those of numerous other non-European civilizations. The existence of numerous other systems is well documented. As an example, the ancient Babylonians had a numeration system based on powers of 60 that still has influence in our society today. There are 60 seconds in a minute and 60 minutes in an hour. The number of degrees in a circle (360) was decided upon by the Babylonians. It is believed they chose this number because it has a lot of divisors so the circle can be divided easily in many ways. The ancient Mayans not only had a system based on powers of 20 but also were one of the first to have a concept and symbol (a seashell) for zero. The ancient Chinese used a series of bamboo sticks (later strokes written in ink) to denote their numbers, and the Egyptians used hieroglyphics. Every civilization we know of had a numeration system, but most people in the Westernized world have learned only about one other than our own. Let that sink in.

Further, the Hindu-Arabic numerals in use today were invented by Indian mathematicians as early as the 1st century. The system was then adopted by Arabic mathematicians in the 9th century. The use of these numerals was promoted by two books: *On the Calculation With Hindu Numerals*, written in 825 by the Indian mathematician al-Khwārizmī, and *On the Use of the Hindu Numerals*, written in 830 by another Indian mathematician, al-Kindī. This numeration system revolutionized how we use numbers, especially in computation. Most texts that introduce numbers and counting do not spend time discussing where the symbols and system came from even though it is a simple enough thing to do even for children. Al-Khwārizmī also wrote about a procedure for solving certain quadratic equations. The procedure, called completing the square, is the basis for the quadratic formula we teach in algebra classes across the United States, and yet often the formula is taught devoid of the history of its development. The title of the book in which this procedure is written, *al-Jabr*, is where our word for algebra comes from. Al-Khwārizmī is often called the father of algebra as he was among the first to consider algebra as a separate area of mathematics though he often justified the procedures he undertook by using geometric explanations. These explanations can and should be shared with our students for they provide a context for where the work came from and a reasoning for why a particular concept works in the first place. This kind of justification is often missing in our mathematics courses.

The most widely recognized theorem is the Pythagorean theorem, which states that the square of the length of the hypotenuse of a right triangle is equal to the sum of the squares of the measures of the other two sides. This theorem is attributed to Pythagoras and has a well-known geometric proof by the Greek mathematician Euclid of Alexandria. However, clay tablets belonging to the Babylonians demonstrate that the Babylonians had extensive knowledge of the relationship expressed in the theorem and did so as early as 1900 BCE, completing complex problems that require its knowledge 2,470 years before Pythagoras was even born. The pyramids of Egypt make clear that the Egyptians had knowledge of the theorem as well. If you stand at the center of the base and look straight up to the point where the sides meet above you, you can trace a perfect 90-degree angle from that point above you straight down the length of your body to the floor beneath your feet and then out to the wall in front of you toward where your toes are pointing. Surely, this did not get built by accident or without knowledge of a theorem central to the workings of right triangles. Another well-known triangle that contains several mathematical patterns and has applications to probability among others is Pascal's triangle. It is named for the French mathematician Blaise Pascal, who was born in 1623. However, like the Pythagorean theorem, it was known to other civilizations centuries prior. The triangle was known in China in the 11th century through the work of a mathematician named Jia Xian. A few centuries later in the 1200s, the mathematician Yang Hui also did work with the triangle, and to this day in China, the triangle is known as Hui's triangle.

Benjamin Banneker

There are many more examples of individuals from marginalized communities contributing to mathematics and yet not having their contributions included in the curriculum, in the texts we rely on, and in the common discourses and stories that make up our social world. Their work and their names are not common knowledge. The stories of these individuals need to be told. They need to be part of the narrative so that those who do not currently see themselves represented in the field begin to realize that people just like them have been there all along.

The *white male math myth* that David Stinson (2013) writes about is so strong that efforts are made to protect the idea that mathematics is the product of Western, white rational thought alone. Benjamin Banneker was born in Maryland to a free Black woman and a formerly enslaved Black man in 1731. He was a Black man who worked as a surveyor, a farmer, and an author, and who had a solid understanding of mathematics. He is often described as self-taught and certainly as someone with mathematical talent and abilities. As an example, he is credited with taking apart and studying a mechanical clock and later building a clock, the first of its kind built in the United States. He is also known for publishing an almanac with extensive information needed for successful farming.

More recently, it has been discovered that his family was from an African community that held extensive knowledge of agriculture and mathematics, which they passed down in oral form one generation to the next. It is likely that Banneker's understanding of mathematics and agriculture came, in part, from the rich tradition of his people. This knowledge was traditionally passed on through the women of certain families, and so it is likely that he learned some of this knowledge from those women to whom it had been taught. Sadly, many times, when a mathematician from a traditionally marginalized community is known, their brilliance in the subject is attributed to their being unique. In this case, the knowledge of Banneker's people and the lengths they took to pass along that knowledge and to protect it have been obscured, and the narrative has been replaced with a story of one who is unique among men. While Banneker was certainly an intelligent and accomplished man, he is the product of a culture that valued mathematical and agricultural knowledge. This knowledge was useful for navigation as well as for farming, yet this rich culture is not celebrated, typically, in the retelling of his story. For more stories of Black mathematicians, consider the text *Beyond Banneker: Black Mathematicians and the Paths to Excellence* (Walker, 2014).

PRIVILEGING SOME STORIES OVER OTHERS

The examples just discussed are only a handful that show that the mathematical contributions of Europeans have been privileged over those of others. Some stories get passed on over and over again. Some people get the credit while others do not. It is time more stories, those of peoples who are not typically placed at the center, are told.

> **It is time more stories, those of peoples who are not typically placed at the center, are told.**

These, too, should be part of the curriculum, as well as part of our popular discourses, and efforts made to erase them should be acknowledged. Incorporating history into the teaching of mathematics gives life to the work being taught. There is, however, still much work to be done around the development of textbooks and classroom materials that incorporate the history of mathematics in a way that values the contributions and life stories of diverse individuals and cultures. Likewise, additional work needs to be done to change the way in which the subject is socially constructed if we are to move past the idea that it is the product, almost exclusively, of Western thought. By including a rich, broad, and diverse history in the teaching of mathematics, we can ensure that we cement more than just the names of white men in our society's knowledge of mathematics.

Similarly, there has recently been a reimagining of education to infuse in it social and political issues with a focus on social justice pedagogy, practices, and curriculum. Though this reimagining is incomplete as traditional instruction remains the norm across the nation, there is resistance to this change from those who claim that it is an attempt to politicize education. To agree that it is such is to deny the reality that education has always been and will always remain political. If the social justice educators and advocates of today are arguing for the prioritizing of the histories, literatures, languages, music, art, and more of those whose stories have been neglected and erased, it is because the current system of education prioritizes other voices instead. Math should be no different. The Eurocentric and male-centered views put forth by texts and traditional curriculum are favoring and privileging whites and, in many cases, white males. This serves to maintain the current and unequal system. It is a political decision to privilege the stories of some over others. We may fail to recognize it because it is what is familiar, what has always been taught, and what has always been done. This, however, does not make it apolitical. The Brazilian educator and activist Paulo Freire (1994), in his groundbreaking text *Pedagogy of the Oppressed*, reminds us that all of education is political:

> There neither is, nor has ever been, an educational practice in zero-space time—neutral in the sense of being committed only to preponderantly abstract, intangible ideas. To try to get people to believe that there is such a thing as this, and to convince or try to convince the incautious that this is the truth, is indisputably a political practice, whereby an effort is made to soften any possible rebelliousness in the part of those to whom injustice is being done. It is as political as any other practice. (pp. 77–78)

Freire argued against what is known as the banking concept of education. Here information (facts, figures, and so forth) is deposited into the minds of students. Instead, he called for students to be fully engaged with the material, seeing education as potentially transformational and a vehicle for liberation and awakening. People, through education, could come to see how the social world is constructed and how they have been positioned within it. This knowledge would then propel them to reclaim society, ensuring that those who have been oppressed would no longer be so through advocacy for a more equitable society. He applied the argument of oppressor and oppressed to education, advocating that education must not serve to perpetuate the patterns of colonialism and oppression present in society.

THE SOCIOPOLITICAL TURN IN MATHEMATICS EDUCATION

Until recently, considering mathematics education as political was not an accepted practice, and for many in the public, the belief that mathematics and mathematics education is apolitical persists. The educational researcher William F. Tate (2005) reminds us, however, that mathematics, too, is political when he writes,

> Until recently, embedding mathematics pedagogy within social and political contexts was not a serious consideration in mathematics education. The act of counting was viewed as a neutral exercise, unconnected to politics or society. Yet when do we ever count just for the sake of counting? Only in school do we count without a social purpose of some kind. Outside of school, mathematics is used to advance or block a particular agenda. (p. 37)

Indeed, mathematics education has taken what Rochelle Gutiérrez (2013) calls a *sociopolitical turn*. By that, she means that the teaching of mathematics is now being used to examine social and political realities. Mathematics is being used by some to fight for social justice while understanding and exploring our social world. What does it mean to teach mathematics for social justice or to highlight the social and political in mathematics? Let us start by considering that mathematics, most especially the contexts we use to teach and apply it, teaches more than just mathematics. As an example, let us consider the following problems:

- Amaya helps her elderly neighbors by doing odd jobs for them throughout the week. If she helped for $\frac{1}{3}$ of an hour every Monday, Wednesday, and Friday and $\frac{1}{2}$ an hour on Saturday and Sunday, how many minutes did she spend helping her neighbors this week?

- It was discovered that a clean engine uses less fuel. An airplane that uses 1,203 gallons of fuel uses 1,189 gallons after it is cleaned for the same trip. If fuel costs $3 a gallon, how much more economical is the clean plane?

The mathematical content covered by the first question is the addition of fractions. The second question focuses on subtraction and multiplication of whole numbers. However, the context of the problems highlights values that are transmitted to students as they work on these. In the first problem, Amaya is helping her elderly neighbors. That we should help others, especially older adults, is being taught by using this context even if the mathematics itself is focused on fractions. In the second problem, the context revolves around the use of fuel and an attempt to save money by using a clean plane. What is being valued here is saving money. The problem could have been framed differently to value something else. For example, instead of focusing on the financial benefit of using the clean plane, a similar problem could have been constructed that focused on how much less pollution is being emitted. In this case, minimizing pollution would have been valued. The way the problem is focused teaches students an important lesson about what is valued.

To bring social issues into the classroom . . .

- Ask students to bring in examples of mathematics in the news, and discuss these as a class.

- Consider replacing the context of problems you use with problems that incorporate social issues or current events.

VALUES TAUGHT BY MATHEMATICS PROBLEMS

In a capitalist society, problems that focus on money and profit are often found in applications of mathematics. The following problem is a feasible region problem where students use the conditions given to maximize a particular variable available at Purplemath's website (purplemath.com/modules/linprog3.htm). Here, we are maximizing profit.

A calculator company produces a scientific calculator and a graphing calculator. Long-term projections indicate an expected demand of at least 100 scientific and 80 graphing calculators each day. Because of limitations on production capacity, no more than 200 scientific and 170 graphing calculators can be made daily. To satisfy a shipping contract, a total of at least 200 calculators much be shipped each day.

> If each scientific calculator sold results in a $2 loss, but each graphing calculator produces a $5 profit, how many of each type should be made daily to maximize net profits?

The problem values technology and has as its goal the aim of maximizing profit. The very same mathematical content can be taught while focusing on a completely different set of values. Take, as an example, the following problem, which is also a feasible region problem.

> The local Union 599 is planning a rally in front of the federal building to "defend" social security. Some people argue that social security dollars should be invested into the stock market or there will not be enough money to support the retired in the future, but the union disagrees. There are two ways to organize for this event: making hour-long blocks of telephone calls and sending out sets of mailings. Including labor, bills, and materials, the cost of a one-hour block of calls is $60 and the cost for one mailing set is $40. Each block of phone calls requires 1 hour, while each set of mailings requires 2 hours to complete. The union can only spend $600 organizing for this demonstration and have agreed [to spend] at least 6 hours coordinating people to attend. They also determined that there should be at least twice as many sets of mailings as hour-blocks of phone calls. Based on past results, every hour worth of calls gets 30 people to come, and every set of mailings gets about 16 people to turn out. Determine what combination of calls and letters will get the most people out to the demonstration. (Osler, n.d.)

The difference in the problems is that the second is focused on maximizing turnout at a rally rather than on maximizing profit. These are put forth here not to value one over the other, per se. In fact, it might make sense to include both when working with students to highlight the applicability of this mathematical content to multiple contexts and to give students more reasons why learning this matters across fields. However, my main goal in highlighting it here is to remind us that mathematics is embedded in context that is value-laden and, as such, political. It also speaks to the values and priorities of those in power in a capitalist society that in mainstream traditional textbooks these problems are typically couched exclusively in language around profits rather than organized labor and demonstrations or any of several viable alternatives.

TEACHING MATHEMATICS FOR SOCIAL JUSTICE

Some mathematics educators have made the decision to use mathematics to promote issues of social justice. An entire pedagogy, the teaching of mathematics for social justice, has developed as a result. Part of what this entails is the use of mathematics to understand and explore issues of social injustice. Rather than examples that are separate from the world around us (for instance, picking marbles out of a jar to explore probability), the examples used to teach mathematics come from our social, political, and economic world. As an example, probability can be explored by relying on data about random police stops where the race of the individual stopped is noted. This is then compared to the population of the area, and a discussion steeped in mathematics can arise as to whether the stops are indeed random. The discussion is, in this way, supported by mathematics as students use mathematics as a tool for exploring social realities and justifying arguments. A second example of embedding social issues into mathematics is to consider the changing face of our representation in government. More women and people of color are being elected in recent elections. Quantifying these changes through mathematics is possible if we consider the percentage increase (or decrease, depending on the case) of women, people of color, and others with marginalized identities to various positions. This may be more engaging to students than determining percentage increase or decrease in problems where the data are fictitious and the context is removed from their experiences. It also provides mathematical language for students to talk about the very society in which they live and the governmental officials that represent them. These examples are taken from the text *Rethinking Mathematics: Teaching Social Justice by the Numbers* by Eric Gutstein and Bob Peterson (2005), who themselves are *math for social justice* advocates and educators and whose work has had much influence in the field. The idea is to use mathematics in ways that highlight the world around us and strive to make that world more just. Student-driven exploration of our social and natural world guides the work, bringing to life the mathematics that is crafted around it.

In this chapter's Resources box you can find a series of texts about math for social justice that highlight the fact that it can be taught at all levels and with numerous topics. These texts include projects that rely on mathematics to explore social issues, such as graphs to explore identity and family composition or the teaching of algebra and functions to explore climate change, culturally relevant income inequality, literacy, and what a just living wage might be. Further examples utilize statistics and probability to explore false positives in drug testing, immigration, the connection between postal codes and test scores, and the prison population. Lastly, geometry is used to explore the availability of healthy food choices by place, the Paralympics, and gerrymandering—the process of manipulating the boundaries of an election district to favor one political party or group of individuals.

A similar text aimed at college-level mathematics, *Mathematics for Social Justice: Resources for the College Classroom* (Karaali & Khadjavi, 2019), uses mathematics as varied as algebra, calculus, differential equations, graph theory, geometry, and discrete mathematics to explore school choice, income inequality, sea level change, the subprime mortgage crisis, student loans, human trafficking, and the rise of acceptance of same-sex relationships. As can be seen, the topics are varied, both in the mathematics used and in the issues addressed. The social issues are timely, relevant, and above all real. These are issues that concern us all, and they are being addressed through mathematics. Such projects challenge students not only to learn rigorous mathematics, but to employ that mathematics to challenge current unjust practices connecting students to their world and positioning them to improve it. We can and should use mathematics to explore the world around us so that students can engage with both mathematics and the world around them. Doing so prepares students to grapple with society's big issues as they move through their schooling and grow into engaged and informed global citizens.

Resources: Teaching Mathematics for Social Justice

Bartell, T. C., Yeh, C., Felton-Koestler, M., & Berry, R. Q., III. (2022). *Upper elementary mathematics lessons to explore, understand, and respond to social injustice*. Corwin.

Berry, R. Q., III, Conway, B. M., IV, Lawler, B. R., & Staley, J. W. (2020). *High school mathematics lessons to explore, understand, and respond to social injustice*. Corwin.

Conway, B. M., IV, Id-Deen, L., Raygoza, M. C., Ruiz, A., Staley, J. W., & Thanheiser, E. (2022). *Middle school mathematics lessons to explore, understand, and respond to social injustice*. Corwin.

Gutstein, E., & Peterson, B. (Eds.). (2005). *Rethinking mathematics: Teaching social justice by the numbers*. Rethinking Schools.

Karaali, G., & Khadjavi, L. S. (Eds.). (2019). *Mathematics for social justice: Resources for the college classroom* (Vol. 60). American Mathematical Society.

Koestler, C., Ward, J., del Rosario Zavala, M., & Bartell, T. (2022). *Early elementary mathematics lessons to explore, understand, and respond to social injustice*. Corwin.

RadicalMath. (n.d.). www.radicalmath.org

HOW CAN YOU CAPITALIZE ON THE POLITICAL NATURE OF MATHEMATICS EDUCATION?

- Consider bringing social issues your students care about into the mathematics classroom.

- Use social contexts related to your students' lived experiences and their communities in the problems you pose to students.

- Value the contributions of diverse people in your classroom.

Questions for Reflection

For Teachers

- How do you value the contributions of diverse cultures in your teaching? What more can you do?

- In what ways can you incorporate social issues into your teaching?

- How might you advocate with your students around social issues that are important to them?

For Instructional Leaders

- In what ways can you support teachers as they navigate the political nature of teaching mathematics?

- How can you support teachers in bringing social issues into their classrooms?

For Administrators

- What opportunities and supports exist for teachers to bring social issues into their teaching?

- How do you support teachers, specifically around communicating with parents and communities, with the incorporation of social justice issues into their teaching?

- How does your school work with parents and community members to foster support and understanding around the incorporation of social justice issues into curriculum and pedagogical practices?

- In what ways can teachers and students be supported in their work as advocates of social change?

- How does the curriculum used by your school engage with social issues?

CHAPTER 11

WHERE DO WE GO FROM HERE?

Throughout this book we have seen numerous examples of socially constructed beliefs about mathematics and mathematics education that have been vigorously challenged. The main reason for the challenge is to push us as a society to rethink the ways in which we engage with the discipline. Our beliefs impact our actions and decisions, which in turn affect the ways in which the subject is taught and learned, the policies that impact our schools and the educational system, and the realities that inform how we move forward in this field. We have seen examples of how policies have driven both access and limits to opportunities. We have seen how beliefs, identities, and issues of justice are intertwined with a discipline that many falsely believe is neutral and apolitical. Until we move past the current set of beliefs that many in society have about mathematics and mathematics education, we will continue to miss opportunities to ensure that many more among us grow to enjoy, appreciate, and understand mathematics.

> Until we move past the current set of beliefs that many in society have about mathematics and mathematics education, we will continue to miss opportunities to ensure that many more among us grow to enjoy, appreciate, and understand mathematics.

Expanding our view of what mathematics entails and who can be successful, shifting the instructional materials and mathematical problems we use, and broadening the ways in which we talk about mathematicians, the stories we tell, the opportunities we offer, and the ways in which we engage and assess students are necessary steps if we are to change a culture that positions women and Black, Latinx, and Indigenous people outside of mathematics while maintaining society's belief that to be *bad at math* is acceptable.

Changing the environment begins by changing our perception of the discipline itself. Until we think of mathematics and mathematics education in a new way, we will continue to marginalize those who do not at present benefit from the way the field is structured. Until we move past the belief that Black and Latinx students do poorly in mathematics, we will continue to offer remediation instead of enrichment and continue to feed the opportunity gaps that exist in the present system, thus perpetuating the very inequities we claim to want to address. Until we realize that the mathematical is political, we will continue to believe that we are supporting all students while centering mathematics instruction on the experiences of the mainstream. Let us begin by pledging to eradicate the belief that it is socially acceptable to admit one is bad at math and move from there, as there is much work to be done.

The work will be difficult. There will be challenges, resistance, and setbacks. It will be easy to slip back into old habits or to think that the problems are too big to address. There are issues raised in this book that go beyond the classroom level, that are structural in nature, and that may seem beyond what we as teachers, instructional leaders, and administrators can transform. It can be discouraging to think about them and feel unable to adequately address them. However, you have more power to do so than you might believe. You can chip away at the issues and the problems a little at a time in a way that seems doable, in a manner that is within your own sphere of influence. I am reminded of the following poem:

> I am only one,
>
> But still I am one.
>
> I cannot do everything,
>
> But still I can do something;
>
> And because I cannot do everything,
>
> I will not refuse to do the something that I can do.

(Grover, 1909, p. 28)

We must choose to make the small changes we as individuals can, even when we cannot yet see a way to make the big changes we, as a society, must.

> ## We must choose to make the small changes we as individuals can, even when we cannot yet see a way to make the big changes we, as a society, must.

There is much work to be done.

This work starts with you. All it takes to get started is one conversation with your students about what mathematics is; one lesson that highlights the mathematics in our everyday lives; one conversation with a parent about what it means to be good at math; one conversation with a teacher about productive struggle or open-ended problems in mathematics; the adoption of one policy to support underrepresented groups in mathematics; the creation and implementation of one alternative assessment module; or the use of one *math for social justice* activity. Each of these steps brings us closer to an equitable system of mathematics education where traditional beliefs are dismantled and no longer hold power to derail our efforts. You have much more power than perhaps you believe you do to impact the field in positive ways. Use that power to effect change.

Consider the various resources listed throughout the book. Reread this book with colleagues, share your reflections and insights, and try some activities using the book study guide found at resources.corwin.com/badatmath. Read as much as you can and raise your voices everywhere that they can be heard. Together we can effect needed change.

> ## Together we can effect needed change.

If the ideas in this book have held meaning for you, consider sharing the book with others and engaging them in conversations around it. You are encouraged to include it in your professional development efforts and in your book club reads. The more we can encourage conversation around these commonly held beliefs, the better able we will be as a society to dismantle them. I'll say it again—it starts with you. Teachers, instructional leaders, and administrators are uniquely positioned to open broad conversations around these issues with students, parents, community members, and policymakers. It is these conversations that will allow us and those we interact with to rethink mathematics education, step outside the usual discourses, and effect substantive change. They have the opportunity to shape the experiences of students in classrooms and teachers in schools in real and substantive ways.

Take a moment now to consider one practice, one step, or one action that you will take toward the goal of dismantling the beliefs that exist around mathematics and mathematics education and effecting positive change. It is this work that will affect the generations of schoolchildren who will come through our school doors—generations who will grow in beliefs that hopefully differ from the way many among us now view mathematics and those who excel at it.

As we have seen, there is much work to do to challenge the social constructs that frame mathematics and mathematics education today, but there are concrete steps we can take. These steps not only help us to push back against the harmful constructs but also help to restructure the discipline, moving it forward in the hopes that with time more among us will learn to enjoy and excel at it. The next time someone admits to you that they are *bad at math*, take the time to ask them why they are comfortable sharing that and whether they would be equally comfortable sharing that they do not know how to read. Challenge them to consider what it means to live in a society that on the one hand churns out reform after reform aiming to strengthen students' knowledge of math and on the other hand accepts that many will not excel in it. Challenge them to think about whether a fundamental shift in this belief might just push us into an era where not a single soul would be comfortable saying they are *bad at math* because, in the end, no one would be.

REFERENCES

Abbott, A. (2014). *The system of professions: An essay on the division of expert labor*. University of Chicago Press.

Adams, J. D. (2021, May 24). Canada should support diversity in STEM to encourage innovative research. *The Conversation*. https://theconversation.com/canada-should-support-diversity-in-stem-to-encourage-innovative-research-146946

Anyon, J. (1980). Social class and the hidden curriculum of work. *Journal of Education, 162*(1), 67–92.

Anyon, J. (1997). *Ghetto schooling: A political economy of urban educational reform*. Teachers College Press.

Bardoe, C., & McClintock, B. (2018). *Nothing stopped Sophie: The story of unshakable mathematician Sophie Germain*. Little, Brown Books for Young Readers.

Baylor, E. (2014). *State disinvestment in higher education has led to an explosion of student loan debt*. Center for American Progress.

Berry, R. Q., III. (2008). Access to upper-level mathematics: The stories of successful African American middle school boys. *Journal for Research in Mathematics Education, 39*(5), 464–488.

Berry, R. Q., III, Conway, B. M., IV, Lawler, B. R., & Staley, J. W. (2020). *High school mathematics lessons to explore, understand, and respond to social injustice*. Corwin.

Bian, L., Leslie, S. J., Murphy, M. C., & Cimpian, A. (2018). Messages about brilliance undermine women's interest in educational and professional opportunities. *Journal of Experimental Social Psychology, 76*, 404–420.

Boaler, J. (2002). *Experiencing school mathematics: Traditional and reform approaches to teaching and their impact on student learning*. Routledge.

Boaler, J. (2015a). *Mathematical mindsets: Unleashing students' potential through creative math, inspiring messages and innovative teaching*. John Wiley & Sons.

Boaler, J. (2015b). *What's math got to do with it? How teachers and parents can transform mathematics learning and inspire success*. Penguin.

Boaler, J. (2022, March 14). Op-ed: How can we make more students fall in love with math? *Los Angeles Times*. https://newsroom.unl.edu/announce/csmce/14455/81534

Boswell, L., & Larson, R. (2010). *Big Ideas math*. Big Ideas Learning.

Bosworth, R. (2014). Class size, class composition, and the distribution of student achievement. *Education Economics, 22*(2), 141–165.

Bourdieu, P., & Passeron, J. C. (1990). *Reproduction in education, society, and culture* (Vol. 4). SAGE.

Bright, A. (2016, Summer). The problem with story problems. *Rethinking Schools*, *30*(4). https://rethinkingschools.org/articles/the-problem-with-story-problems/

Brock, L. (2020, October 7). Using a book club to navigate challenging topics. *Edutopia*. https://www.edutopia.org/article/using-book-club-navigate-challenging-topics

Brown, M. (Director). (2015). *The man who knew infinity* [Film]. Pressman Film, Xeitgeist Entertainment Group, & Cayenne Pepper Productions.

Bruine de Bruin, W., Parker, A. M., & Fischhoff, B. (2007). Individual differences in adult decision-making competence. *Journal of Personality and Social Psychology*, *92*(5), 938–956.

Bruno, R. (2018, June 20). When did the U.S. stop seeing teachers as professionals? *Harvard Business Review*. https://hbr.org/2018/06/when-did-the-u-s-stop-seeing-teachers-as-professionals

Burdman, P. (2022, March 15). To keep students in STEM fields, let's weed out the weed-out math classes. *Scientific American*. https://www.scientificamerican.com/article/to-keep-students-in-stem-fields-lets-weed-out-the-weed-out-math-classes/

Cai, J. (2021, August 1). Safe and healthy buildings: Infrastructure problems revealed by the Nation's Report Card. *American School Board Journal*. https://www.nsba.org/ASBJ/2021/august/safe-and-healthy-buildings

Card, D., & Giuliano, L. (2016). Universal screening increases the representation of low-income and minority students in gifted education. *Proceedings of the National Academy of Sciences of the United States of America*, *113*(48), 13678–13683.

Carraher, T. N., Carraher, D. W., & Schliemann, A. D. (1985). Mathematics in the streets and in schools. *British Journal of Developmental Psychology*, *3*(1), 21–29.

Carver-Thomas, D., & Darling-Hammond, L. (2017). *Teacher turnover: Why it matters and what we can do about it*. Learning Policy Institute.

Chaabane, S., Doraiswamy, S., Chaabna, K., Mamtani, R., & Cheema, S. (2021). The impact of COVID-19 school closure on child and adolescent health: A rapid systematic review. *Children*, *8*(5), 415. https://doi.org/10.3390/children8050415

Chan, P. C. W., Handler, T., & Frenette, M. (2021). Gender differences in STEM enrolment and graduation: What are the roles of academic performance and preparation? *Statistics Canada*, *1*(11), 1–19.

Chval, K. B., Smith, E., Trigos-Carrillo, L., & Pinnow, R. J. (2021). *Teaching math to multilingual students, Grades K–8: Positioning English learners for success*. Corwin.

Cohen, P. C. (2003). Democracy and the numerate citizen: Quantitative literacy in historical perspective. In B. L. Madison & L. Steen (Eds.), *Quantitative literacy: Why numeracy matters for schools and colleges* (pp. 7–20). National Council on Education and the Disciplines.

Common Core State Standards Initiative. (2021). *Mathematics Standards*. http://www.corestandards.org

Complete College America. (2021). *Data dashboard*. https://completecollege.org/data-dashboard/

Coogler, R. (Director). (2018). *Black panther* [Film]. Marvel Studios.

Dee, T., & West, M. (2011). The non-cognitive returns to class size. *Educational Evaluation and Policy Analysis, 33*(1), 23–46.

Després, S., Kuhn, S., Ngurimpatse, P., & Parent, M. (2014). *Real accountability or illusion of success? A call to review standardized testing in Ontario.* Action Canada Task Force on Standardized Testing.

Devlin, K. (2005). *The math instinct: Why you're a mathematical genius (along with lobsters, birds, cats, and dogs).* Thunder's Mouth Press.

Devlin, K. (2012). *Introduction to mathematical thinking.* Keith Devlin.

DiAngelo, R. (2018). *White fragility: Why it's so hard for white people to talk about racism.* Beacon Press.

DiAngelo, R. (2021). *Nice racism: How progressive white people perpetuate racial harm.* Beacon Press.

Douglas, D., & Attewell, P. (2017). School mathematics as gatekeeper. *The Sociological Quarterly, 58*(4), 648–669.

Elawar, M. C., & Corno, L. (1985). A factorial experiment in teachers' written feedback on student homework: Changing teacher behavior a little rather than a lot. *Journal of Educational Psychology, 77*(2), 162–173.

Faulkner, V. N., Marshall, P. L., & Stiff, L. V. (2019). *The stories we tell: Math, race, bias, and opportunity.* Rowman & Littlefield.

Feiner, L. (2020, April 5). How NYC moved the country's largest school district online during the coronavirus pandemic. *CNBC.* https://www.cnbc.com/2020/04/03/how-nyc-public-schools-are-shifting-online-during-the-coronavirus.html

Feldman, J. (2019). *Grading for equity: What it is, why it matters, and how it can transform schools and classrooms.* Corwin.

Ferguson, E. (2018, December 7). Parents worry students could be squeezed out of overcrowded schools in their own communities. *Calgary Herald.* https://calgaryherald.com/news/local-news/parents-worry-students-could-be-squeezed-out-of-overcrowded-schools-in-their-own-communities

Forgasz, H. J., & Leder, G. C. (2017). Persistent gender inequities in mathematics achievement and expectations in Australia, Canada, and the UK. *Mathematics Education Research Journal, 29*(3), 261–282.

Freire, P. (1994). *Pedagogy of the oppressed* (3rd ed.). Continuum.

Fry, H. (2015). *The mathematics of love: Patterns, proofs, and the search for the ultimate equation.* Simon & Schuster.

Gallegos, E. (2022, May 17). In the San Joaquin Valley, rapidly growing school districts endure overcrowding. *EdSource.* https://edsource.org/2022/in-the-san-joaquin-valley-rapidly-growing-school-districts-endure-overcrowding/672175

García, E., & Weiss, E. (2020). How teachers view their own professional status: A snapshot. *Phi Delta Kappan, 101*(6), 14–18.

Gerber, C., & Mitchell, J. (Executive producers). (2016–2020). *Elena of Avalor* [TV series]. Disney Television Animation.

Gonzalez, L. (2018). Between paralysis and empowerment: Action in mathematics for social justice work. *New England Mathematics Journal, LI*(2), 33–40.

Gonzalez, L., Chapman Brown, S., & Battle, J. (2020, December). Mathematics identity and achievement among Black students. *High School Science and Mathematics, 120*(8), 456–466. https://doi.org/10.1111/ssm.12436

Gonzalez, L., Lucas, N., & Battle, J. (2022, May 7). A quantitative study of mathematics identity and achievement among Latinx secondary school students. *Journal of Latinos and Education.* Advance online publication. https://doi.org/10.1080/15348431.2022.2073231

Gordon, M., Davis, J., Bernero E. A., Spera, D., Mundy, C., Mirren, S., Messer, E., Barrois, J. S., Frazier, B., Bring, H., & Kershaw, G. (2005–2020). *Criminal minds* [TV series]. The Mark Gordon Company, Entertainment One, Touchstone Television, Bernero Productions, ABC Studios, Paramount Network Television, CBS Television Studios.

Grover, E. O. (1909). *The book of good cheer: A little bundle of cheery thoughts.* PF Volland.

Gutiérrez, R. (2008). A gap-gazing fetish in mathematics education? Problematizing research on the achievement gap. *Journal for Research in Mathematics Education, 39*(4), 357–364.

Gutiérrez, R. (2013). The sociopolitical turn in mathematics education. *Journal for Research in Mathematics Education, 44*(1), 37–68.

Gutiérrez, R. (2017). Why mathematics education was late to the backlash party: The need for a revolution. *Journal for Urban Mathematics Education, 10*(2), 8–24.

Gutstein, E., & Peterson, B. (Eds.). (2005). *Rethinking mathematics: Teaching social justice by the numbers.* Rethinking Schools.

Hargreaves, A. (2000). Four ages of professionalism and professional learning. *Teachers and Teaching, 6*(2), 151–182.

Harrison, D. (2020–present). *Kids math talk* [Audio podcast]. https://www.kidsmathtalk.com/podcast

Hottinger, S. N. (2016). *Inventing the mathematician: Gender, race, and our cultural understanding of mathematics.* SUNY Press.

Ingersoll, R. (2003, September 1). *Is there really a teacher shortage?* University of Pennsylvania Graduate School of Education. https://repository.upenn.edu/gse_pubs/133

Ingersoll, R. M., & Collins G. J. (2018). The status of teaching as a profession. In J. Ballantyne, J. Spade, & J. Stuber (Eds.), *Schools and society: A sociological approach to education* (6th ed., pp. 199–213). Pine Forge Press/SAGE.

Inglis, M., & Attridge, N. (2017). *Does mathematical study develop logical thinking? Testing the theory of formal discipline.* World Scientific.

Jackson, C. K., Wigger, C., & Xiong, H. (2021). Do school spending cuts matter? Evidence from the great recession. *American Economic Journal: Economic Policy, 13*(2), 304–335.

Jaschick, S. (2019, February 11). More AP success; racial gaps remain. *Inside Higher Ed.* https://www.insidehighered.com/admissions/article/2019/02/11/more-students-earn-3-advanced-placement-exams-racial-gaps-remain

REFERENCES

Johnson, J. J., McKeon, T., Powers, B., Siefken, P., Bishop, M. J. R., & Lopez, G. (2014–present). *Odd squad* [TV series]. Sinking Ship Entertainment & Fred Rogers Productions.

Karaali, G., & Khadjavi, L. S. (Eds.). (2019). *Mathematics for social justice: Resources for the college classroom* (Vol. 60). American Mathematical Society.

Keenan, C., McCorkindale, S., & Pistor, J. (2018–present). *Barbie dreamhouse adventures* [TV series]. Mainframe Studios & Mattel Television.

Kempf, A. (2016). *The pedagogy of standardized testing: The radical impacts of educational standardization in the US and Canada.* Springer.

Kobett, B. M., & Karp, K. S. (2020). *Strengths-based teaching and learning in mathematics: Five teaching turnarounds for Grades K–6.* Corwin.

Konstantopoulos, S., & Chung, V. (2009). What are the long-term effects of small classes on the achievement gap? Evidence from the lasting benefits study. *American Journal of Education, 116*(1), 125–154.

Kozol, J. (2005). *The shame of the nation: The restoration of apartheid schooling in America.* Crown.

Kraft, M. A. (n.d.). *Understanding teacher turnover.* https://f.hubspotusercontent20.net/hubfs/2914128/Upbeat_Literature%20Review.pdf

Lazarín, M. (2014). *Testing overload in America's schools.* Center for American Progress.

Learning Forward. (2013). *Tool: Creating norms.* https://learningforward.org/lf-newsletter/summer-2013-vol-8-no-4-4/tool-creating-norms/

Leslie, S. J., Cimpian, A., Meyer, M., & Freeland, E. (2015). Expectations of brilliance underlie gender distributions across academic disciplines. *Science, 347*(6219), 262–265.

Liljedahl, P. (2020). *Building thinking classrooms in mathematics, Grades K–12: 14 teaching practices for enhancing learning.* Corwin.

Loomis, E. S. (1968). *The Pythagorean proposition.* Classics in Mathematics Education Series.

Lorre, C., Prady, B., Aronsohn, L., Kaplan, E., Ferrari, M., & Goetsc, D. (2007–2019). *The big bang theory* [TV series]. Chuck Lorre Productions & Warner Brothers Television.

Luzniak, C., & Baier, R. (2022). *DebateMath* [Audio podcast]. https://debatemath.com/

Malkevitch, J. (2008, April). *The process of electing a president.* American Mathematical Society Feature Column. http://www.ams.org/publicoutreach/feature-column/fcarc-elections

Martin, D. B. (2009). Does race matter? *Teaching Children Mathematics, 16*(3), 134–139.

McKellar, D. (2007). *Math doesn't suck: How to survive middle school math without losing your mind or breaking a nail.* Hudson Street Press.

McKellar, D. (2009). *Kiss my math: Showing pre-algebra who's boss.* Hudson Street Press.

McKellar, D. (2010). *Hot X: Algebra exposed.* Hudson Street Press.

McKellar, D. (2012). *Girls get curves: Geometry takes shape.* Hudson Street Press.

Melfi, T. (Director). (2016). *Hidden figures* [Film]. Fox 2000 Pictures, Chernin Entertainment & Levantine Films.

Mendivil, A., & Alcorn, M. (2021–present). *Sum of it all* [Audio podcast]. Anchor by Spotify. https://anchor.fm/sdcoe-math

Merton, R. K. (1948). The self-fulfilling prophecy. *The Antioch Review, 8*(2), 193–210.

Meyer, D. (2010, March). *Math class needs a makeover* [Video]. TED Conferences. https://www .ted.com/talks/dan_meyer_math_class_needs_a_makeover

Meyer, M., Cimpian, A., & Leslie, S. J. (2015). Women are underrepresented in fields where success is believed to require brilliance. *Frontiers in Psychology, 6*, 235. https://doi.org/10.3389/ fpsyg.2015.00235

Mosca, J. F., & Rieley, D. (2018). *The girl with a mind for math: The story of Raye Montague.* Innovation Press.

Moser, J. S., Schroder, H. S., Heeter, C., Moran, T. P., & Lee, Y. H. (2011). Mind your errors: Evidence for a neural mechanism linking growth mind-set to adaptive posterror adjustments. *Psychological Science, 22*(12), 1484–1489. https://doi.org/10.1177/0956797611419520

Moses, R. P., & Cobb, C. E. (2001). *Radical equations.* Beacon.

National Center for Education Statistics. (2019, February). *Status and trends in the education of racial and ethnic groups: Indicator 23: Postsecondary graduation rates.* U.S. Department of Education. https://nces.ed.gov/programs/raceindicators/indicator_red.asp

National Center for Science and Engineering Statistics. (2019). *Women, minorities, and persons with disabilities in science and engineering: 2019.* Special Report NSF 19-304. National Science Foundation. https://ncses.nsf.gov/pubs/nsf19304/

National Center for Science and Engineering Statistics. (2021, April 29). *Women, minorities, and persons with disabilities in science and engineering: 2021.* Special Report NSF 21-321. National Science Foundation. https://ncses.nsf.gov/pubs/nsf21321/

National Council of Teachers of Mathematics. (2013). *Matching the Common Core State Standards for Mathematics (CCSSM) and Canadian content expectations, content outcomes, and Essential Knowledges for Ontario, Québec, or Western and Northern Canadian Protocol curriculums.* https://www.nctm.org/canadiancorrelations/

National Council of Teachers of Mathematics. (n.d.). *The role of elementary mathematics specialists in the teaching and learning of mathematics.* https://www.nctm.org/Standards- and-Positions/Position-Statements/The-Role-of-Elementary-Mathematics-Specialists-in-the- Teaching-and-Learning-of-Mathematics/

Ontario Ministry of Education. (2021, December 8). *New math curriculum for Grades 1–8.* https://www.ontario.ca/page/new-math-curriculum-grades-1-8

Organisation for Economic Co-operation and Development. (2019). *TALIS 2018 results (Vol. 1): Teachers and school leaders as learners.* OECD Publishing.

Osler, J. (n.d.). *Organizing for the future: Lesson plans.* RadicalMath. https://www.radicalmath .org/news-1/organizing-for-the-future

REFERENCES

Partelow, L., Shapiro, S., McDaniels, A., & Brown, C. (2018). *Fixing chronic disinvestment in K–12 schools.* Center for American Progress.

Pasquale, M. (2016). *Productive struggle in mathematics: Interactive STEM Research+ Practice Brief.* Education Development Center.

Pearce, K., & Orr, J. (2018–present). *Make math moments that matter* [Audio podcast]. Make Math Moments. https://makemathmoments.com/podcast/

Pulfrey, C., Buchs, C., & Butera, F. (2011). Why grades engender performance-avoidance goals: The mediating role of autonomous motivation. *Journal of Educational Psychology, 103*(3), 683.

Purvis, L. (2021, February 25). COVID-19 school closures and the National School Lunch Program. *Georgetown Journal on Poverty, Law and Policy.* https://www.law.georgetown.edu/poverty-journal/blog/covid-19-school-closures-and-the-national-school-lunch-program/#_edn13

RadicalMath. (2007). www.radicalmath.org

Rattan, A., Good, C., & Dweck, C. S. (2012). "It's ok—Not everyone can be good at math": Instructors with an entity theory comfort (and demotivate) students. *Journal of Experimental Social Psychology, 48*(3), 731–737.

Rattan, A., Savani, K., Naidu, N. V. R., & Dweck, C. S. (2012). Can everyone become highly intelligent? Cultural differences in and societal consequences of beliefs about the universal potential for intelligence. *Journal of Personality and Social Psychology, 103*(5), 787–803.

Reid, M. (2021). *Maryam's magic: The story of mathematician Maryam Mirzakhani.* HarperCollins.

Schecter, S., Gero, M., Pellington, M., Berlanti, G., & Siega, M. (2015–2020). *Blindspot* [TV series]. Berlanti Productions, Quinn's House & Warner Brothers Television.

School Nutrition Association. (2020). *Impact of COVID-19 on school nutrition programs: Part 2.* https://schoolnutrition.org/uploadedFiles/11COVID-19/3_Webinar_Series_and_Other_Resources/COVID-19-Impact-on-School-Nutrition-Programs-Part2.pdf

Shapiro, D., Dundar, A., Huie, F., Wakhungu, P. K., Bhimdiwala, A., & Wilson, S. E. (2018, December 18). *Completing college: A national view of student completion rates—Fall 2012 cohort* (Signature Report No. 16). National Student Clearinghouse Research Center. https://nscresearchcenter.org/signaturereport16/

Shetterly, M. L. (2016). *Hidden figures: The American Dream and the untold story of the Black women who helped win the space race.* William Morrow.

Shetterly, M. L. (with Conkling, W.), & Freeman, L. (Illustrator). (2018). *Hidden figures: The true story of four Black women and the space race.* HarperCollins.

Singh, S. (2021). *Chasing rabbits: A curious guide to a lifetime of mathematical wellness.* IM Press.

Snider, S., Burk, D., & Smith, T. (1999). *Bridges in mathematics 2: A Math Learning Center curriculum.* Math Learning Center.

Stanford, L. (2022, August 10). How many teachers are retiring or quitting? Not as many as you might think. *Education Week.* https://www.edweek.org/leadership/how-many-teachers-are-retiring-or-quitting-not-as-much-as-you-might-think/2022/08?utm_source=nl&utm_medium=eml&utm_campaign=eu&M=4850806&UUID=a8a841704681bc77ce4971e71a34a711

Steele, C. M. (2006). Stereotype threat and African-American student achievement. In D. Grusky & S. Szwlwnyi (Eds.), *The inequity reader* (pp. 252–257). Westview Press.

Stephens-Davidowitz, S. (2014, January 18). Google, tell me. Is my son a genius? *The New York Times*. https://www.nytimes.com/2014/01/19/opinion/sunday/google-tell-me-is-my-son-a-genius.html

Stewart, J., Redlin, L., & Watson, S. (2005). *Precalculus: Mathematics for calculus* (5th ed.). Brooks/Cole.

Stigler, J. W., & Hiebert, J. (2009). *The teaching gap: Best ideas from the world's teachers for improving education in the classroom*. Simon & Schuster.

Stinson, D. W. (2013). Negotiating the "white male math myth": African American male students and success in school mathematics. *Journal for Research in Mathematics Education, 44*(1), 69–99.

Tate, W. F. (2005). Race, retrenchment, and reform mathematics. In E. Gutstein & B. Peterson (Eds.), *Rethinking mathematics: Teaching social justice by the numbers* (pp. 31–40). Rethinking Schools.

Tsekouras, P. (2021, September 9). Some Ontario parents say there are 30 to 40 students in their children's classroom. *CTV News*. https://toronto.ctvnews.ca/some-ontario-parents-say-there-are-30-to-40-students-in-their-children-s-classroom-1.5578961

U.S. Department of Education Office for Civil Rights. (2018, April). *2015–16 civil rights data collection STEM course taking: Data highlights on science, technology, engineering, and mathematics course taking in our nation's public schools*. https://ocrdata.ed.gov/assets/downloads/stem-course-taking.pdf

Vasquez Heilig, J., Williams, A., & Jez, S. (2010). Inputs and student achievement: An analysis of Latina/o-serving urban elementary schools. *Association of Mexican American Educators Journal, 10*(1), 48–58.

Walker, E. N. (2014). *Beyond Banneker: Black mathematicians and the paths to excellence*. SUNY Press.

Wang, K., Rathbun, A., & Musu, L. (2019, September). School choice in the United States: 2019 (NCES 2019-106). U.S. Department of Education, National Center for Education Statistics. https://nces.ed.gov/pubsearch/pubsinfo.asp?pubid=2019106

Williams, S. W. (2000, December 12). *Black women in mathematics: 3 African American women × 3 Ph.D.s = one rare achievement in mathematics* [Press release]. Mathematicians of the African Diaspora. State University of New York at Buffalo Mathematics Department. http://www.math.buffalo.edu/mad/special/3X3.html

Woodrow, D., Rose, D., Braun, S., O'Brien, W., Kadin, H., Lin, J., Wootton, N., Santora, N., Kurtzman, A., Orci, R., & Foster, D. (Executive Producers). (2014–2018). *Scorpion* [TV series]. K/O Paper Products, Black Jack Films, Perfect Storm, SB Projects, & CBS Television Studios.

Wynn, K. (1992). Addition and subtraction by human infants. *Nature, 358*, 749–750.

Zeichner, K. (2020). Preparing teachers as democratic professionals. *Action in Teacher Education, 42*(1), 38–48.

Zhang, J., Kuusisto, E., & Tirri, K. (2017). How teachers' and students' mindsets in learning have been studied: Research findings on mindset and academic achievement. *Psychology, 8*(9), 1363–1377. https://10.4236/psych.2017.89089

INDEX

**JENNIFER M. BAY-WILLIAMS,
JOHN J. SANGIOVANNI,
ROSALBA SERRANO,
SHERRI MARTINIE,
JENNIFER SUH, C. DAVID WALTERS**

Because fluency is so much more
than basic facts and algorithms.
Grades K–8

**ROBERT Q. BERRY III, BASIL M. CONWAY IV,
BRIAN R. LAWLER, JOHN W. STALEY,
COURTNEY KOESTLER, JENNIFER WARD,
MARIA DEL ROSARIO ZAVALA,
TONYA GAU BARTELL, CATHERY YEH,
MATHEW FELTON-KOESTLER,
LATEEFAH ID-DEEN,
MARY CANDACE RAYGOZA,
AMANDA RUIZ, EVA THANHEISER**

Learn to plan instruction that engages
students in mathematics explorations
through age-appropriate and culturally
relevant social justice topics.

**Early Elementary, Upper Elementary,
Middle School, High School**

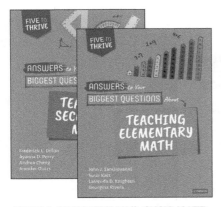

**JOHN J. SANGIOVANNI, SUSIE KATT,
LATRENDA D. KNIGHTEN,
GEORGINA RIVERA,
FREDERICK L. DILLON,
AYANNA D. PERRY,
ANDREA CHENG, JENNIFER OUTZS**

Actionable answers to your most
pressing questions about teaching
elementary and secondary math.

Elementary, Secondary

**SARA DELANO MOORE,
KIMBERLY RIMBEY**

A journey toward making
manipulatives meaningful.
Grades K–3, 4–8

A SAGE Publishing Company

Helping educators make the greatest impact

CORWIN HAS ONE MISSION: to enhance education through intentional professional learning.

We build long-term relationships with our authors, educators, clients, and associations who partner with us to develop and continuously improve the best evidence-based practices that establish and support lifelong learning.